D1058437

COSMIC
DISCOVERIES

FURTHER PRAISE FOR
COSMIC DISCOVERIES

"This is a fascinating account of a spectrum of astronomical discoveries and the people who made them. The book sparkles with excitement and insight. Levy himself has experienced the magic of cosmic discoveries, and has personally known several of the astronomers of whom he writes."

—Roy L. Bishop, Ph.D., past president,
The Royal Astronomical Society of Canada

"David and Wendee Levy describe the thrill of discovery as only astronomers who have 'been there, done that' really can."

—Paul Weissman, Senior Research Scientist, Jet Propulsion Lab

"Science is much more than a logical search for cold facts; it is a passionate activity pursued for the joy of discovery. David Levy knows this from his own experience, and expresses that joy better than anyone."

—Stuart Weidenschilling, Ph.D., Senior Scientist,
The Planetary Science Institute

The Wonders of Astronomy

COSMIC

DISCOVERIES

DAVID H. LEVY

with WENDEE WALLACH-LEVY

 **Prometheus Books**

59 John Glenn Drive
Amherst, New York 14228-2197

Published 2001 by Prometheus Books

Inquiries should be addressed to
Prometheus Books
59 John Glenn Drive
Amherst, New York 14228–2197
VOICE: 716–691–0133, ext. 207
FAX: 716–564–2711
WWW.PROMETHEUSBOOKS.COM

05 04 03 02 01   5 4 3 2 1

Library of Congress Cataloging-in-Publication Data

Levy, David H., 1948-
   Cosmic discoveries : the wonders of astronomy /
David H. Levy with Wendee Wallach-Levy.
      p. cm.
   Includes index.
   ISBN 1–57392–931–X
   1. Astronomers—Biography 2. Astronomy—History. I. Wallach-Levy,
Wendee. II. Title.
QB35 .L48 2001
520'.92'2—dc21
[B]   2001048280
                                                           CIP

Printed in the United States of America on acid-free paper

for

Nanette and Mark

and our granddaughter, Summer

May your lives be filled with the joy of discovery

ACKNOWLEDGMENTS

In an undertaking such as this, it is not possible, nor is it appropriate, to include a chapter on everyone who has made an important discovery in astronomy. Choosing which stories to include was a challenge, and we thank our large group of friends and associates who helped us make our selection.

To Linda Regan of Prometheus Books, we owe a profound debt; this is our fifth book together. In preparing this book, we have received assistance from many people, including Joan-ellen and Philip Rosenthal, Nanette Vigil, Tim Hunter, Don McCarthy, Dan Green, Carolyn Shoemaker, Jean Mueller, James Reston Jr., Steve O'Meara, Clyde Tombaugh, Bart Bok, Geoffrey Marcy, Paul Butler, Robert Summerfield, and Sir Arthur C. Clarke.

We thank *Sky & Telescope* magazine for permission to use material from David Levy's "Star Trails" column, especially for chapters 21 and 22. The Astronomical Society of the Pacific has helped with providing some of the photographs of people we have used in this book.

We also want to thank our many friends and family who encouraged us to complete this book. In particular, Roy Bishop, a physicist and a dear friend from David's alma mater, Acadia University, went through every sentence of the manuscript and offered invaluable suggestions. He also pointed out that Dale Frail, who co-discovered three extrasolar planets (see chapter 20), is a graduate of Acadia University. Way to go, Acadia!

Finally, while on the way to train for a swimming competition, we both decided that we did not have the time to complete this book and therefore would not proceed with it. An hour-long swim has a marvelous effect in clearing the mind. Near the end of our laps, David called over to Wendee and said, "This book is really worth doing!" We left the pool energized and excited. Were it not for that swim, there probably would be no book. Thank you, Charles Wacker from City of Tucson Parks and Recreation!

CONTENTS

FOREWORD

As you go about your daily life, you never know if some strange event will take place that will catapult you into the history books. What seems like just a normal routine to some may actually be considered an unimaginable feat by others. In the pages to come, we describe the lives and achievements of a selection of the greatest men and women in astronomical history. Some of these people are already familiar to you, while others are still carving their niche in history.

While doing the research for each chapter, David and I had long discussions about what each of these discoverers must have been like. For example, when I thought of Galileo, the picture of an old man in a stone house came to mind. When I was a girl, I thought he did little more than just look at the sky and make discoveries. Now I understand that there is so much more to the man—his stubbornness, his perseverance, his courage, and his greatness.

Two character traits that I found in all these people are dedication and stubborn persistence. They are all quite single-minded and do not waver from their goals until they are met. Galileo wrote, studied, and taught even while near death and imprisoned in his home. The dethroned president of the Paris Parliament, Jean-Baptiste-Gaspard de Saron, was still doing orbital calculations for Charles Messier while lying terrified in prison awaiting the guillotine!

The process of discovery, we've been told, is 1 percent inspiration and 99 percent perspiration. We've tried to expand not only on "the perspiration process," but the inspirational sparks, and those aspects of everyday life that added to the process as well. Who would ever expect a great astronomer to be so cavalier as to have his nose cut off in a bar fight! Meet Tycho Brahe. Then there was Clyde Tombaugh who, one night while guiding his telescope for an hour-long exposure, found himself getting strangely sleepy. When the picture ended, he couldn't move his fingers and he found he was on the verge of hypothermia. With tremendous difficulty he was able to close the telescope, and return to a heated room, but not before he put the photographic plate safely in its darkroom. Near-death experiences aren't a necessary part of the discovery process, but they do show the commitment that it requires.

As you read each chapter, I hope you get as caught up in the thrill and excitement of the discovery process as we did. David has lived through the thrill of discovery twenty-one times. Each night when David and I go into the observatory to search for comets, I hope that soon I, too, will have that thrill.

Because of the different natures of the discovery stories, some chapters offer more background material on the discoverers than others. We took into account the specific details and personalities in making our decisions on how much personal material to include in each chapter. A final note: Even though David and I have co-authored this book, we have decided on a specific style to use whenever a personal story involving one of us is told. A story using first person ("When I first met Clyde Tombaugh") refers to David. A story involving me ("Wendee joined our comet hunting team") is in third person.

Welcome to the world of "How they did what they did," and how it feels to make a discovery. As your mind travels to the times and places offered in the following pages, remember that all these people saw, observed, recorded, and discovered a sky that has hardly changed at all. May your own path be lit by these same stars.

Wendee Wallach-Levy

PREFACE

YEAR 2000:
AN ANCIENT COMET
RETURNS

W hen the alarm rang at 3 A.M. on the morning of August 9, 2000, the predawn sky was clear, dark, and filled with stars. Even though I was tired, the summer had been so cloudy that the sight of stars brought a rush of adrenalin. Within a few minutes I was awake, dressed, out the back door, and on my way to the far end of the yard.

There, a hundred yards or so from our home, lies Jarnac Observatory, which was named after the cottage my grandfather owned in the Gatineau Hills, northwest of Montreal, the site of some of my earliest views of a dark sky. Jarnac is my temple. Jarnac was built in two stages. The first part consists of a twelve-foot-square structure with a roof that slides off, on wheels, and the second is a large, new addition, 14-by-20 feet, that has a roof that also rolls away. When both parts are open, several telescopes can peer to the sky.

On that clear August morning, I opened only the older, smaller section. I unfastened the chains, and gave the roof a shove. Obediently, almost silently, its eight coaster wheels moved the roof to the north, revealing the whole sky above me. Next, I turned on the power to Miranda, my sixteen-inch telescope. Named after the Roman goddess of wisdom, Miranda is also the young woman who, in Shakespeare's *Tempest*, cries out "O brave new world, That has such people in't." I felt that Miranda was an appropriate name for a telescope that has so far discovered seven new comets.

This sixteen-inch telescope actually uses very little electric power. Part of its electricity runs a beautiful little device that, once set up, tells me exactly where in the sky the telescope is pointing. The rest of the power is for my father's small

13

radio, which hangs on the telescope near the eyepiece, providing music and fond memories of times I've had with Dad.

Once it was set up, I swung the telescope to the northeast and started comet hunting. I looked at the telescope's first field of view, an area of sky to photograph. For just a second or so I examined the field to see if any fuzzy spots lurked among the stars. Comets appear as hazy patches of light, but a "faint fuzzy" is far more likely to be a distant cluster of stars, a nebula, or a distant galaxy than a comet. This field showed no such fuzzies, so I moved on to the next . . . and the next . . . moving the telescope up the sky a few fields . . . then over a field . . . then down a few . . . until I had covered a significant patch of the heavens and found nothing unusual. Then I swung Miranda back to the north again. In the half hour of searching, new stars had risen, and I was now looking at a patch of sky near Pollux and Castor, the two bright stars of the constellation Gemini, the twins.

Not far from these two bright stars, I stopped the motion of the telescope. In the field of view was a faint fuzzy spot! I checked the position, and quickly determined from my electronic pointer that no galaxy or other known fuzzy object was supposed to be there. I made a sketch of the position, then went on searching for a while. Ten minutes later, when I returned to the site of the suspect, my heart leaped. The fuzzy spot had moved!

I had found a comet, but was it a new comet? It was time to head indoors, turn on the computer, and log into a special program of the International Astronomical Union's Central Bureau for Astronomical Telegrams. I typed in the celestial coordinates of the comet, a procedure I had done several times before. While the computer, located at Harvard's Center for Astrophysics, did its work, I remembered other occasions when it came back with "no objects found"—a pretty good sign that the comet I had located was new. This time, I held my breath.

After what seemed like eternity but which really was less than a minute, the computer came back with the positions for one of the oldest known comets. With a sudden letdown of tension, I learned that I had just "discovered" Encke's Comet.

Ah, what could have been! But disappointment very quickly gave way to a feeling of immense satisfaction. For although my comet was a well-known visitor, its return in 2000 was one of the worst since its discovery more than two hundred years ago. Throughout the few months that it could conceivably be seen, it was so deep in twilight that observation was made very difficult. More important, I had located one of the most famous comets in history. If during my years of comet searching I had to stumble on an already known comet, Encke is the one I would want to find.

What a history this comet has! On January 17, 1786, French comet hunter Pierre Méchain found a comet of fifth magnitude, meaning it was barely visible to the naked eye. Nine years later, on November 7, 1795, famous English comet hunter Caroline Herschel discovered the comet, now fainter at magnitude 5.5, as it returned. The sister of William Herschel, probably England's most famous astronomer, could not know that she had picked up Méchain's comet.

On October 20, 1805, another French comet hunter, Jean-Louis Pons, discovered it for a third time, also at magnitude 5.5. By this time comet hunting was quite a popular activity, and Eugene Bouvard independently found this comet shortly after Pons. Meanwhile, when a German mathematician named Johann Franz Encke calculated an orbit for this comet, he was surprised to suggest that it might be a comet that returns every twelve years.

About half a century earlier, Halley's comet had proved itself to be the first such returning, or "periodic," comet, and here was the possibility of a second one. On November 26, 1818, Pons once again discovered this comet. Encke then connected Pons's new comet to the 1805 comet, but in so doing he realized that its period was not twelve years, but three and a third years. Encke also calculated that Pons's comet was the same as those of Méchain and Herschel, confirming the orbit of this singular comet: No other known comet orbits the Sun as quickly as does Comet Encke.

Although Encke always believed the comet should be named for Pons, who discovered it twice, it is now named Comet Enke in honor of the mathematician who figured out its strange orbit. On June 2, 1822, the Australian searcher Carl Rumker recovered this comet, confirming Encke's work once and for all. Since then it has been followed on most of its returns. When I found Encke's Comet on August 9, 2000, it was about one hundred times fainter than those earlier discoveries. For me, on that morning, I felt as though the door of history had opened.

DISCOVERY IN ASTRONOMY

The book you are about to read is about discovery in astronomy. Comets, and the people who found them, are an important part of that story: Edmond Halley's returning comet in 1759, a quarter-century before the story of Encke began, showed that comets are a part of our solar family. When Halley's Comet returned in 1986, astronomers observed its supply of carbon, hydrogen, oxygen, and nitrogen—the building blocks of life itself. But it was not until the discovery of Comet Shoemaker-Levy 9 in 1993 and its subsequent impact with Jupiter a year later that humanity gained a real understanding of how comets have affected the development of life on Earth. Comets don't just carry the fundamentals of life; they deposit those building blocks during collisions with the planets, including Earth.

Through its stories of men and women who have peered through telescopes and found new worlds, this book traces the discovery adventure through the major "finds" that took place through the centuries. When someone looks through a telescope and discovers something new—a comet, a planet, or, like

Encke, the confirmation of an idea—we expand our understanding of the neighborhood of which we on Earth are a part. Some discoveries are especially important in this regard. Tycho Brahe's supernova in 1572 planted the seeds of a new understanding of the universe. A generation later, Galileo Galilei's discovery of the moons of Jupiter, new worlds that refused to orbit the Earth, challenged the whole doctrine of the Earth being the center of the universe. With the start of the twentieth century, Harlow Shapley pushed back the envelope that had been opened by Galileo—if the Earth was not the center of the universe, neither was the Sun, for the center of our galaxy was very far away. Before the first quarter of this century was over, Edwin Hubble showed that even our galaxy is but a tiny part of a universe that is expanding.

Fewer than four hundred years have passed since Galileo's small telescope pointed out the changing nature of the world around us. In that time we have seen the universe expand, and our solar system become a fearful place, with comets colliding with planets. Thanks to these discoveries, and the advent of wondrous technologies, we stand on the verge of being able to understand our corner of the universe just a little bit better.

Will the next stage of astronomical discovery merge science with technology in a fruitful way? Will it involve a comet headed for Earth, and our ability to send a spacecraft out to change its course? This book's final chapter will explore this remote but real possibility. The thought of this happening brings us back to Encke's comet, which wandered into astronomical record books just as our ways of thinking about the sky were being fundamentally changed. In another way, however, discovery confirms that old French adage, *plus ça change, plus c'est la même chose* (the more things change, the more they stay the same). When I open my observatory in the predawn sky, the centuries fall away, and I feel as though I am standing beside Méchain, Pons, and Herschel, all of us wondering what surprise the sky has in store for us.

ONE

1842: TO SAIL BEYOND THE SUNSET

Portraits of Discovery

> *Come, my friends,*
> *'Tis not too late to seek a newer world.*
> *Push off, and sitting well in order smite*
> *The sounding furrows; for my purpose holds*
> *To sail beyond the sunset, and the baths*
> *Of all the western stars, until I die.*
> *It may be that the gulfs will wash us down;*
> *It may be we shall touch the Happy Isles,*
> *And see the great Achilles, whom we knew.*
> *Tho' much is taken, much abides; and tho'*
> *We are not now that strength which in old days*
> *Moved earth and heaven; that which we are, we are;*
> *One equal temper of heroic hearts,*
> *Made weak by time and fate, but strong in will*
> *To strive, to seek, to find, and not to yield.*
> —Alfred, Lord Tennyson, *Ulysses*, 1842[1]

> *There it was that I found and visited the famous Galileo, grown old, a pris-*
> *oner to the Inquisition for thinking in Astronomy otherwise than the Fran-*
> *ciscan and Dominican licensers of thought.*
> —John Milton, *Areopagitica*, 1644[2]

The Shoemaker-Levy Double Cometograph is a complex name for a simple and beautiful instrument of discovery. It is the result of a dream that began one night as Gene and Carolyn Shoemaker and I were partaking in our

17

comet and asteroid search program at Palomar Observatory, California. This program, which had resulted in the discoveries of many new worlds, including the great Comet Shoemaker-Levy 9 that extinguished its life in a spectacular collision with Jupiter, was about to end. Late in 1994, during one of our last observing sessions with that project, Gene and I were taking a photograph of the star fields, while Carolyn, one floor below, was scanning the films that we had taken earlier.

"We need to find a way for Carolyn to keep finding comets," Gene began as we marked time for a photographic exposure that would last eight minutes. By the end of the exposure, we had agreed on a plan for a new program using two wide-field telescopes known as Schmidt cameras. Gene called them cometographs, in honor of their intended purpose to record comets.

Within a year, our new two-camera program was already being tested. Like any search program, Gene was delighted that this new project was getting off the ground and producing consistent searches of the night sky for comets. In the years to come, the Shoemaker-Levy Double Cometograph project would continue to build on Gene's prescription: Cover as much sky as you can, keep good records and statistics, take advantage of breaks in the clouds, and never, never give up.

As I write this, Tennyson's adage of "To strive, to seek, to find, and not to yield" seems true in all except the "find" part. Though Gene is no longer with us, his hope and dream is very much alive in our program each evening as Carolyn, my wife, Wendee, and I begin our photographic search. Our films show regions of the sky, taken over and over again, but no new comets have been captured by them as yet. We have expanded the program recently with a wonderful new twelve-inch Schmidt camera from Meade Instruments Corporation. Also, we have joined forces with Carol Neese and Gil Esquedero of the Planetary Science Institute in Tucson to add an electronic component to our search. An electronic camera—a CCD, or charge-coupled-device—has been attached to one of our telescopes, and we hope that this novel technique will aid our search for comets in the near-Sun areas through which I have looked for so many years.

SEARCHING

This book, however, is a *direct* result of the Shoemaker-Levy Cometograph program. Our search program involves taking two exposures of each field of sky each night. To accomplish this, we take a series of several photographs of different fields of sky. When the series is complete, we repeat it. The films are then prepared for scanning and finally they are placed into a device called a stereomicroscope.

A few months ago, I placed two films into the stereomicroscope and began searching them for comets. The two films were of an identical area of the sky, but were taken the previous evening about forty-five minutes apart. As the stars passed before my eyes, I felt myself taking my place in line with other men and women who had done the same thing, either with their own eyes, by watching through a telescope, looking carefully at pictures, or in modern times scanning an electronic image. It was at that moment that I got the idea to write this book.

Why do people search the sky? What's in it for them? And more important, what's in it for the rest of us? Of these questions, the answer to the last one is easiest. Were it not for those who, like Ulysses, sailed beyond the sunset, our knowledge and understanding of the universe would not be anywhere near where it is today. We might still be wandering the halls of ancient universities, proud of our ability to weave a perfect argument, and oblivious of whether that argument really explains how the universe works. We are indebted to the great discoverers of history, for they have forced us to look not at the universe as it seemed, but as their new findings show it to be.

There is no unique and simple answer to the question of why people search the sky. Here we will explore the discoveries and lives of some of the people who made them in an effort to shed light on this question. One of the most remarkable discoverers was Galileo Galilei. He spent the early part of his professional life doing experiments that included dropping objects of different weights from the Leaning Tower of Pisa, and creating the first liquid thermometer by showing that the density of a fluid varies inversely with its temperature. Had he done only these things, he would be remembered as one of a number of important scientists of the Renaissance. Until 1595, for example, he was one of many who accepted Ptolemy's view that the Earth was the center of the universe, despite Copernicus's alternate idea, that the Sun was the center, which had been proposed decades before. In that year, however, the thirty-one-year-old scientist determined that the Earth's rotation on its own axis, plus its revolution around the Sun, offered a better explanation of the tides than did Ptolemy's view.

Galileo's fertile mind, combined with a confluence of historical events, pushed him to go far beyond the tides. His great contribution arose from his building a telescope, looking through it, measuring with it, and thinking carefully about the results he obtained with it. The change came in 1609, when Galileo heard of the invention of a new device that could magnify distant objects. By 1610 Galileo had recognized the awesome potential of this "spyglass." Increasing its power, he turned it toward the sky, and discovered that Jupiter had four moons that clearly revolved about the giant planet. He knew that Jupiter's moons were strong evidence, but not proof, that the Earth was not the center of the universe. When Venus appeared in the sky later that year, he used his telescope to discover that it had phases that simply would not be seen in that way if Venus orbited Earth. Galileo could explain his own observations of the changing phases of Venus *only* if Venus orbited the Sun.[3]

Two years later, Galileo developed a means of predicting the times when Jupiter's moons would pass behind Jupiter. Again, these predictions were accurate only when he took into account the changing position of his observing site on Earth as it orbited the Sun. Over a two-year period of observation, Galileo's telescope provided him with incontrovertible proof that the planets orbit the Sun, not the Earth. With the combination of telescope and mind, Galileo left humanity a new vision of the Earth's place, the first one in about two thousand years.

MILTON'S VISIT WITH GALILEO

Now we move forward from 1612 to 1638. Almost thirty years had passed since Galileo first lifted his telescope to the sky, and much had changed. In the spring of 1633 he had faced the Inquisition in Rome, and was forced to retract his statements about the Earth orbiting the Sun, even though they had been confirmed by his telescope. Now near seventy years of age and blind, he was sentenced to life imprisonment even though he had been led to believe that a far lighter sentence was in store. Galileo was so stunned by this harsh sentence that he nearly died. Subsequently the sentence was lightened; Galileo was forced to remain for the rest of his life in his villa in Florence, under guard by the Office of the Inquisition.

Five years later, a young English poet named John Milton was traveling through Switzerland and Italy. In September 1638 Milton visited Galileo. Milton's brief sentence about the visit, quoted at the start of this chapter, was written in his 1643 treatise *Areopagitica: A Speech for the Liberty of Unlicensed Printing to the Parliament of England*. Milton's treatise was political, but the poet was so moved by the strength of mind he perceived in the aged scientist in Florence that he never forgot that day. Twenty-three years after Galileo's death in 1642, John Milton completed his own great contribution to humanity, his epic poem *Paradise Lost*. In it he mentions Galileo twice, once even by name:

> The broad circumference
> Hung on his shoulders like the moon, whose orb
> Through optic glass the Tuscan artist views
> At evening, from the top of Fesolé,
> Or in Valdarno, to descry new lands,
> Rivers, or mountains, in her spotty globe.[4]

In another example, Milton compares the lens of Galileo to the eye of the "Eternal Father," who sees everything clearly; Galileo is "less assured" as he uses his telescope to study the Moon:

>   . . . as when by night the glass
> Of Galileo, less assured, observes
> Imagined lands and regions in the Moon;
> Or pilot from amidst the Cyclades
> Delos or Samos first appearing kens,
> A cloudy spot.[5]

Can it be said that the discoveries of a scientist were of such magnitude to stimulate the mind of one of the greatest English poets? Milton left his meeting with Galileo troubled that such a man would lie imprisoned in his home merely because of his ideas. That meeting is an important symbol of the compelling influence of science on literature, and of the great power of astronomical discovery that affects all of us. The nineteenth-century English writer Walter Savage Landor guessed at what their conversation might have entailed. "Let us talk of something else," Galileo says when Milton brings up the subject of torture at the hands of the Inquisition. To which Milton said angrily, "Italy, Italy, Italy! Drive thy poets into exile, into prison, into madness. Spare thy one philosopher! What track can the mind pursue, in her elevations or her plains or her recesses, without the dogging and prowling of the priesthood?"[6]

THE MIND OF THE DISCOVERER

In one of his early books, *The Exploration of Space*, Arthur C. Clarke told the story of Pheidias, the sculptor who spent years creating the frieze on the Parthenon. Would someone have asked him, Clarke wrote, "why he was not engaged in something useful like rebuilding the Athenian slums[?] If he had kept his temper, the artist would probably have answered that he was doing the only job that interested him. So it is, in the ultimate analysis, with those who want to cross space."[7] People who discover, I contend, do it because it is the one job that interests them, and they cross space without even leaving the Earth.

"You have to have the imagination," Clyde Tombaugh told me, "to recognize a discovery when you see one." As the discoverer of Pluto, the solar system's most distant major planet, Tombaugh was in a good position to know this. "When they examined *Voyager* images," he went on about the intrepid spacecraft traveling through the outer solar system, "and saw for the first time the volcanic eruptions on Io, that called for some intuitive imagination."

"I would suggest," Tombaugh continued, "that above everything else, in observing you have to be very alert to everything. You have to be able to recognize a discovery as such. There are so many people who don't seem to have that talent. A research astronomer cannot afford to be in such a rut. I might say that different types of personalities in astronomy make certain types of discoveries that are in line with their personalities."[8]

For people like Galileo, who discovered the moons of Jupiter, and Clyde Tombaugh, who discovered Pluto, the type of thinking that shapes their younger years is likely to produce a person with the will to sail beyond the sunset. It is incorrect to say that such people are driven by others, but it is correct to say that they drive themselves. Neither Galileo nor Tombaugh set goals to discover worlds, but they both carefully monitored the goals they did set, and they both were aware of what could lie on the road to these goals. Take Tombaugh, a man I admired for many years. Despite the fact that I wrote his biography, it was not until long after that book was out of print and Tombaugh was gone that I finally got the chance to see some of the goals he had set for himself as a young man. The list which follows was found by his wife, Patsy, when she was going through his papers a few years after his death. It is a looking glass that gives us an insight into the determination of the man. Tombaugh began the list no earlier than 1921, when he was fifteen years old, and kept it, on a double-sided sheet of lined paper, for at least five years:

THINGS TO DO[9]

*continuous

No.	Things to do— projects	Origin of Intention (Time)	When Completed
1.	Make a lantern slide box	March, 1923	Cancelled June 21, 1925
2.	Read *Pilgrim's Progress*	April, 1923	Cancelled June 9, 1925
3.	Study part 1 of *Sanitation and Physiology*	Oct. 1924	Completed Aug. 10, 1925
4.	Work the problems about star rising in the back of Trigon. book	Dec. 1923	
5.	Reread "Julius Caesar"	Feb. 1924	Cancelled June 21, 1925
* 6.	Study Spanish	Sept. 1921	college
7.	Read "Revelations" of the Bible	Aug. 1924	Completed Dec. 1924
* 8.	Print astronomical notes	1920	—
9.	Buy Chemistry book and study it	March 1924	Cancelled September 10, 1925
10.	Reread "Story of the Solar System"	August 1924	Completed June 21, 1925.

11. Measure height of east hill from lower dam by the 2L method	April 1924	Cancelled June 21, 1925
12. Finish reading Psalms	Dec. 9, 1923	Completed Feb. 1924
13. Reread my Ancient History	Sept, 1924	Canceled June 14, 1925
14. Find Pickering's Station in Jamaica	Dec. 1923	
15. Make a glass vacuum tank	Oct. 1924	Canceled May 11, 1925
16. Finish football field	April 1924	
17. Study Esther's "*Modern Progress*"	Oct. 1922	Completed thoroughly on July 24, 1926.
18. Read a Kansas History	Jan. 1925	Half read, done with it June 12, 192?
*19. Paint and draw pictures	1922	
20. Compare areas and volumes of spheres.	May 1924	college
21. Make transit instrument	March 1924	Cancelled and completed Jan. 6, 1926
22. Make opera glass attachment to tripod	Sept. 1924	Completed September 30, 192?
23. Make photographic plate camera	Oct. 1924	Cancelled Feb. 14, 1926
24. Study Dad's "Zoology"	Jan. 1925	Canceled June 14, 1925
25. Study Dad's "Physics"	Jan. 1925	Completed Aug. 23, 1925
26. Study Dad's Chemistry	Jan. 1925	Cancelled (college)
27. Make a toy steam turbine engine	Jan. 1925	Cancelled March 26, 1926
28. Make a Leyden jar	Jan. 1925	Canceled May 11, 1925
29. Make a static electrical machine	Jan. 1925	Canceled May 11, 1925
30. Make a water decomposing apparatus	Feb. 1925	Canceled March 26, 192?
31. Buy and fit a prism attachment for telescope	Oct. 1922	Higher magnification not possible August 11, 1925.

32. Make a 10 [crossed out] 8" reflecting telescope	Dec. 1924	
33. Take a trip to Illinois	1923	Completed Dec. 28, 1925.
34. Take a trip to Flagstaff Arizona to look thru the big telescope.	July, 1924	
35. Finish reading "Ecclesiastes"	Feb. 16, 1925	Completed March 12, 1925
36. Study solid geometry	Feb. 16, 1925	College
37. Finish Plane geometry notebook	Dec. 1924	Cancelled
38. Buy Mary E. Byrd's "Laboratory Manual of Astronomy" R. 9, Box 77, Lawrence, Kans.	March 5, 1925	
39. Finish Studying Phy. Geo. [Phys. Geography]	May 20, 1925	Completed July 4, 1925
40. Finish Studying U.S. Constitution	May 20, 1925	Completed June 14, 1925
41. Finish Studying Com. Law		June 9, 1925
42. Study lunar craters in "Astronomy with an opera glass"	June 21, 1925	
43. Fix generator and make use of it.	June 28, 1925	
44. Put up roof-cooling apparatus	June 28, 1925	Cancelled, March 27, 1926
45. Compose a song of lunar scenery to the tune of "America the Beautiful"	Oct. 20, 1925	Completed Jan. 6, 1926
46. Buy accordion and learn to play it.	Dec. 26, 1925	

Clyde Tombaugh was an extremely focused man. I was delighted with this list, which says so much about Tombaugh's interests as a youth, but I wasn't surprised by it. We will read more about both Galileo and Tombaugh in later chapters. Our purpose here is to open the subject of what kind of person tends to become a discoverer. For more than twenty years after his discoveries of 1610, Galileo wrote, studied, and tried to ensure that the public understood the significance of what he had found. In that he succeeded beyond measure, but at a tremendous personal cost. No discoverer in astronomy approached the seriousness of the consequences of Galileo's findings with his telescope. In Tombaugh's case, although the discovery of Pluto opened opportunities for him, it also aroused jealousies that caused him problems in later years.

Both Galileo and Tombaugh, however, focused their lives in unusual ways. In Clyde Tombaugh's table, for example, lay the essence of the mind of a discoverer. Although he completed just over a quarter of the projects he started, even the completed ones show an unusually eclectic range of interests; less than a quarter of the projects are astronomy-related. The management of the projects themselves, and the maintenance of the list, seemed to be of intense interest to him as well. Each project was considered, planned, and either carried to fruition or canceled. The original sheet of paper must have been in active use for several years before Tombaugh stored it away. During the seventeen years I knew him, neither he nor Patsy ever mentioned it, and Patsy was very surprised to find it in 1999. As I went through the list, I tried to anticipate which items he cancelled (no. 24: Study Dad's Geology was one), and which he completed (no. 25: Study Dad's Physics).

Two of the items on this list have made their way into history. No. 32, which began as a ten-inch-diameter and ended up as an eight-inch reflecting telescope, was completed four years later, despite the missing date, but its quality was poor. Tombaugh followed this telescope by building a nine-inch reflector of superb optical quality. Two items later, No. 34, called for a trip to Flagstaff to "look thru the big telescope." Again, no completion date was recorded. Tombaugh sent several samples of drawings he had made using the nine-inch telescope, it turned out, to the director of the Lowell Observatory in Flagstaff, a mailing that began a chain of events that resulted in his heading for Flagstaff in 1928.

In a real sense, item No. 34 should have been marked "Completed" on February 20, 1930. On that evening, he and two members of the observatory's senior staff walked up to the dome for the twenty-four-inch refractor, took the covers off the old tube and looked "thru the big telescope." At the other end was the planet, later to be named Pluto, that Tombaugh had discovered two days before on February 18, 1930.

TWO

1965: PASSPORT FOR DISCOVERY

Time has not lessened the age-old allure of the comets. In some ways, their mystery has only deepened with the years. At each return a comet brings with it the questions which were asked when it was here before, and as it rounds the sun and backs away toward the long, slow night of its aphelion it leaves behind with us those questions, still unanswered.

To hunt a speck of moving haze may seem a strange pursuit, but even though we fail the search is still rewarding, for in no better way can we come face to face, night after night, with such a wealth of riches as old Croesus never dreamed of.

—Leslie C. Peltier, 1965[1]

I first got the idea for a comet search program in October 1965. The astronomy world was abuzz that month with the promise of a truly great comet that had been found just two weeks earlier by two Japanese amateur astronomers, Kaoru Ikeya and Tsutomu Seki. Both young men had discovered comets before, but this one, which they found independently in the predawn sky while searching in the constellation of Hydra, was special. Shortly after its discovery, astronomers realized that the comet's orbit was carrying it directly toward the Sun. By the third week of October, the comet would pass within 200,000 miles of the Sun's surface and would be bright enough to be seen in daylight.

My young, seventeen-year-old mind thought what a wonderful time this must be for those two amateur astronomers. Here was a comet whose presence they had revealed to the world weeks before, and now they would be watching as the comet drew its majestic course across the sky. I couldn't imagine anything

26

more exciting than that. I was in tenth grade, at Westmount High School in Montreal, Canada. That October our French teacher had scheduled a series of early morning oral exams, so each Wednesday morning I would walk the mile to school. I was excited about the comet, but during these walks I also had to think about my answers to the questions on my French oral exam. I knew that one of them would relate to my future—what do you want to have as a career?

Suddenly the thought hit me. *"Je veux decouvrir une comete!"* I want to discover a comet. I thought about how much fun it would be to explain, *en Français,* about comets and their orbits, about how they are discovered, and about the careers of the two intrepid Japanese comet hunters Ikeya and Seki. The plan worked; during French orals we had a wonderful discussion about comets and astronomy. The only problem occurred when the teacher asked if people could actually earn a living finding comets. They can't, of course; comet hunting was then strictly a pursuit for amateur astronomers. For a living, I told them about the two things I've wanted to do: be an astronomer and a writer.

On other Wednesday morning walks to school, I thought about the possibilities for my new plan to search for comets. I realized that the chances for me actually finding a comet were extremely remote. Only five or six new comets bright enough to be discovered through amateur telescopes visit the inner part of the solar system each year, and I thought that many hundreds of amateur astronomers would be competing to find them. But although the year 1965 was in its fall, my life was in its spring and I had lots of time for dreams. Although I doubted that I would ever discover a comet of my own, I grew excited about the prospect of simply searching the sky for comets. The search, I thought, would be its own reward.

METHODS OF ASTRONOMICAL DISCOVERY THROUGH HISTORY

My decision to begin comet hunting took place at a time when discovery methods were changing. The tradition of visually finding new objects dated back to ancient times, when careful observers would look up and see a "guest star" in the sky. Probably the best known visual discovery in history is that of Danish astronomer Tycho Brahe: He saw and studied a bright new star in the constellation of Cassiopeia in 1572. Naked eye observers have the entire sky at their disposal; the great advantage of the unassisted human eye is that it can cover more sky at one time than any other device. A man who knew the sky very well, Tycho would have found that bright new star anywhere in the sky, probably within minutes! Not even the fastest modern CCD detector can do that well. The disadvantage of a lack of light-gathering power offered by a telescope is that the human eye cannot see stars fainter than about sixth magnitude, perhaps seventh if the sky is dark and the eyes well trained. Even with such training, finding a comet or a nova fainter than about third magnitude with the naked eye is doubtful. (Megrez, the star that joins the handle with the bowl of the Big Dipper, is magnitude 3.3.) Still, each time I go outdoors at night, I do give the sky a quick look without any optical aid, just to see if anything new has happened since the night before. (The magnitude scale is a way of measuring the brightness of an object in the sky. The higher an object's magnitude, the fainter it is and therefore the more difficult to see.)

The minute Galileo pointed his telescope to the heavens, he upped the ante. His discovery of the moons of Jupiter began a series of visual discoveries with telescopes that continues to this day. On March 23, 1850, William Cranch Bond opened the new world of astrophotography by exposing the first good "daguerrotype" photograph of the Moon. Almost half a century would pass, however, before photography would reveal a new comet. Edward Emerson Barnard found Periodic Comet Barnard 3 as a trailed smudge on a photographic plate in 1892.

By 1965, more than half the comets were being found on photographic plates, so when I began, I knew that my competition was both from other amateur astronomers as well as from professionals searching the sky. By 1981, electronics added a new method of surveying for comets. That year a satellite called *Solwind 1* became the first spacecraft to discover a comet electronically when it found one on a collision course with the Sun. The several *Solwind* satellites, the Solar Maximum Mission, the *IRAS* satellite, and especially the *SOHO* satellite detected many comets. Near the end of 1990, Comet Spacewatch became the first comet found on an image taken with a CCD chip. Today, more comets are found electronically than by any other means. In the world of professional astronomy, film is hardly used any more; virtually all discoveries of all manner of cosmic objects are made by CCDs.

CN-3 BEGINS!

This new program of comet searching, I was certain, would face many hurdles. As early as 1966, a proposed military program to search for enemy satellites threatened to scan the entire sky, every night, for comets. With all this competition I considered, from the beginning, the ultimate success of my search program quite doubtful. *But it was my program*, my own statement of a sky search for comets and exploding stars, or novae. I even had a code name for it: CN-3, for Comet and Nova Search 3. (CN-1 was a simpler project using binoculars, and CN-2 was specifically a program to observe Comet Ikeya-Seki in 1965.) Finally, during a break in the clouds near midnight on Friday evening, December 17, 1965, I, at age seventeen, conducted my first comet search between the bright stars Pollux and Castor in Gemini. Energized by that brief first encounter with my search program, I went inside and carefully inscribed its three aims:

1. To become very familiar with the sky through searching for comets and/or novae.
2. To discover either a comet or a nova.
3. To learn as much as possible about comets and/or novae through a research program.

"As of Dec. 17, 1965," I wrote also, just in case anyone would question that later, "the main interest area is in the field of comets."[2]

After an hour or so of observing and writing, I realized that my program to search the sky for comets had begun. I was now officially a night watchman, whose responsibility it would be to search the sky for a new comet.

There were other searches going on. At the end of 1967 Ikeya and Seki shared a second comet, and the following year Minoru Honda, one of Japan's most successful comet hunters, bagged a nice one. One night, while hunting outside my parents' home I made an independent discovery of Comet Honda. I realized it had to be a known comet; and it did not take long to find out that it was Honda's. My grandmother, who was visiting at the time, shared my excitement as I darted in and out of the house between star charts and telescope. It didn't matter that the comet was already known; I had found a comet. For me, it was a night of discovery.

A year later I was far from home. Under the dark sky at Acadia University in Nova Scotia, I set up my telescope. In an effort to get away from the lights of campus, I walked a mile—uphill—with telescope, mount, and notebook in hand. It was worth the walk, though in retrospect I wish I still had that kind of energy. I set up the tripod that evening, placed the telescope on it, inserted an eyepiece in the telescope, and looked through it. Near the top edge of the field was a comet! It turned out to be a known one, a comet that a Japanese amateur, Osamu Abe, had discovered a few months earlier. Another evening of discovery.

Life got in the way after that, and although my comet hunting continued, there were no other "independent discoveries," as finds of known comets are called. In the late summer of 1977 I independently discovered a nova—an exploding star—in the constellation of Cygnus, just a few days after it had been first reported. In the spring of 1979 I took a long look at the project. CN-3 had now gone on for fourteen years, and although I enjoyed it very much, I wanted to bring closure to it. If I really wanted to raise my chances of finding a new comet, I thought I must relocate to a place with a more consistently clear sky. Thus, in August 1979 I moved to the southwest desert of Arizona for the specific purpose of successfully completing my comet search. Even so, it wasn't until April 1983, after 834 hours of searching, that I found yet another known comet. After a flurry of excitement, I checked the pages of *Sky & Telescope* magazine and learned that my new comet was actually a well-known old one, Periodic Comet Tempel 2. Then at the end of November I found Comet Hartley-IRAS only a few days after its discovery! After 863 hours, I was hoping my time was near.

FIRST COMET

The afternoon of Tuesday, November 13, 1984, was cloudy, but a clearing sky at sunset made me cut short a dinner date and rush home to begin my comet search. As was my usual practice, I moved Miranda, my sixteen-inch reflector telescope, from field to field. After some thirty minutes had gone by a faint fuzzy object appeared in Miranda's field of view. It had the appearance of a galaxy, I thought, and a quick check of an atlas confirmed my suspicion. The next object was a planetary nebula, the remnant of an outburst in an ancient star. I stopped to enjoy this wonder in the night, but not for long: like Robert Frost, I had miles to go before I could sleep.

I continued until my clock indicated that I had completed an hour of comet hunting. Normally I would stop at that point, but the sky was so warm and pleasant that I decided to keep going. Three minutes later I encountered NGC 6009, a pretty cluster of stars. But my attention was drawn to a fuzzy object in the same field of view, a bit to the south. The sight of cluster and fuzzy object was so striking that I wondered why I had never seen it before. When my star atlas confirmed the star cluster, but not the fuzzy object, I began to get really excited. Out came my sketch pad. I drew the cluster, a few field stars, and the fuzzy patch. When I checked the field a quarter hour later, I was sure that the object was moving very slowly in the direction of the cluster. I was in heaven, out in space with my first comet!

It was a comet all right, but was it already known? I spoke over the telephone with Brian Skiff, an observer at the Lowell Observatory some 300 miles away in Flagstaff. The five minutes before his return call seemed like an eternity. When the phone rang again I grabbed the receiver with both hands. "You'd better send a telegram," Brian said, "You've got a comet." After 917 hours, 28 minutes, spread over nineteen years, my search was over.

THREE RAPID-FIRE DISCOVERIES,
AND THE START OF A FRIENDSHIP

Or was it? I had a feeling of absolute satisfaction and relief as I telegrammed Brian Marsden, director of the International Astronomical Union's Central Bureau for Astronomical Telegrams, with the news of the new comet. This single taste of success, however, merely whet my appetite for more. I resumed comet hunting only nine days later, and soon found my second comet, a faint, elongated patch of fuzzy light, on the morning of January 5, 1987. I didn't have to wait long for my third. On October 12, 1987, while testing a new observing site atop the roof of my home, I discovered my third comet only 107 observing hours after the second. Nor did I have to wait long for a fourth; on March 20, 1988, I found a comet that began a strange story and a wonderful friendship.

That story began early in March 1988, at a conference in Tucson about aster-oids. That's where I first met Gene and Carolyn Shoemaker, two geologists who had begun their own comet search program a few years earlier. While my program's only scientific purpose was to learn as much as possible about comets, theirs had a specific aim: to gather statistics about the numbers of comets and asteroids that could pose a threat to the Earth. I was fascinated by the new area of study and importance that they had given to comets, now seen as harbingers of destruction. Comets are not just interesting for what they are, but for what they can do.

As if to cement our new friendship, two weeks later I discovered Comet Levy 1988e. On hearing of my find, Gene and Carolyn decided to add it to their observing program at Palomar. At the end of their night, they swung the telescope over to the east and took a brief exposure of the field that contained the new comet. Despite the brightening dawn sky they got a good image and submitted the first accurate positions of my comet.

The next month Gene and Carolyn, now with colleague Henry Holt, again included my comet on their list of fields of sky to photograph. They set up their list graphically on a sheet of paper on which nickel-sized circles outline the observing fields around the sky. But the comet was north and east of where they usually photograph, so Gene placed an extra circle at the top of his diagram.

On the morning of May 13, 1988, the telescope was pointed at the position indicated by the extra circle. The following evening, Carolyn placed the Comet Levy films on her stereomicroscope and quickly found a comet. Gene looked also, and wondered why the comet was so far from the field's center. Carolyn then tried to establish the comet's position relative to the surrounding stars, and found that the stars on the field did not match the field they were supposed to be photographing. They had photographed the field represented by the position of the circle on the diagram, instead of the field where Comet Levy was. They had not photographed Comet Levy that night, but instead had discovered a new comet!

After enough positions of the new comet, Shoemaker-Holt 1988g, allowed a calculation of the interloper's orbit, Conrad Bardwell of the Central Bureau for Astronomical Telegrams made a discovery of his own. The orbits of Comet Levy and Comet Shoemaker-Holt, he noticed, were almost identical in every respect except that Comet Shoemaker-Holt arrived at its closest point to the Sun, or perihelion, some three months after Comet Levy's closest approach to the Sun. This was a singular instance of a pair of related long-period comets being discovered independently. The two comets were one some 12,000 years ago which, for some reason, split apart.[3] They are continuing their separate journeys around the Sun, moving away from each other, and will be years apart when they next return.

A NEW DECADE OF COMETS

August 26, 1989, was indeed a night for discovery. The spacecraft *Voyager 2* was completing its tour of the outer solar system by photographing Neptune and its great moon Triton. Even though August in Tucson, Arizona, is known for its poor weather, that night was clear, so I alternated between watching each new photograph of distant Triton appear on my television set, and heading outdoors to search for comets. After an hour and a quarter of on-again, off-again searching, my views of Triton were completely interrupted by my discovery of a new comet. This one I shared with the Japanese amateur astronomer Okazaki and my friend Michael Rudenko, who was conducting a comet search from Massachusetts.

JOINING THE SHOEMAKER TEAM

By the fall of 1989, I was a part of Gene and Carolyn Shoemaker's observing program using the eighteen-inch Schmidt camera at Palomar Mountain. Their program represented a whole new way of observing for me. Instead of looking through a telescope to search for comets, in this new survey a team of observers took photographs of areas of the sky and then scanned these photographs, a pair at a time. Any object, like a comet or an asteroid, that changed its position from one film to the next would appear, through the stereomicroscope, to float above or below the background of stars.

CN-3 FINALLY TURNS TO NOVAE

As originally set up, my search program included novae, or exploding stars, as well as comets. I had the opportunity to find such a star, but in a strange way.

While looking through the old plate archives at Lowell Observatory as part of research for my biography of Pluto discoverer Clyde Tombaugh, I found an interesting entry on the envelope of a plate taken on March 23, 1931: It claimed that Tombaugh had detected a possible nova in Corvus. "Evidently a very remarkable star," he wrote, "to rise from 17 or fainter to 12 in 2 days time." The discovery was never announced; the nova was buried in old emulsion.

Intrigued by this note, I wanted to try to confirm Tombaugh's discovery. But doing so for a nova that appeared in the sky almost sixty years earlier meant that my "observing" would not be in the sky, but on old photographic plates stored at the Harvard College Observatory in Cambridge, Massachusetts. Disappointed at first, I found no confirmation of the outburst in 1931. Since I had only an hour before someone was supposed to meet me, I passed the time by looking at a sampling of more recent plates of the region of the constellation of Corvus, where Tombaugh had found the star. On the tenth plate I searched, I noticed the star was exploding again! That plate had been taken during the 1970s. This star, it seemed, had a history of outbursts. Energized by this find, I checked through all 360 patrol plates of Corvus in the Harvard collection, and discovered nine additional outbursts of this star. That fall, I began a series of observations of the star's field to see if I could catch it visually in a new outburst. For almost seventy nights I looked. Then, on the night of March 23, 1990, I pointed Miranda toward the proper field in Corvus. There was the star Tombaugh had found exactly fifty-nine years to the day before; I was witnessing it explode, visually, for the first time. Now named TV Corvi, the star is Tombaugh's discovery. My role was to take the evidence of the discovery and complete the work needed to announce the find to the world. Fifteen months later, the *International Ultraviolet Explorer Satellite* studied this star. To see the star rescued from being a speck on a photographic plate to being the subject of investigation by a space satellite left me deeply thrilled.

. . . AND BACK TO A BRIGHT COMET . . .

Despite the activity at Palomar, and the sighting of the nova, I kept up my own visual observing program. This program paid off two months after I observed Tombaugh's star when, on May 20, 1990, I found Comet Number 6 moving its careful way among the stars of Pegasus. It was faint the night of discovery, but over the next few months it brightened rapidly as it drew closer to both Earth and Sun. Throughout the late summer of 1990, Comet Levy (1990c) was visible around the world as it cruised along the Milky Way.

. . . AND AN ASTEROID TURNS INTO A COMET

Meanwhile, our observing at Palomar was proceeding well. In November 1990 the first Shoemaker-Levy comet shed a few photons of light onto our films. The previous month we had found an asteroid that had a most unusual orbit that took it out toward Jupiter's distance from the Sun and back again in a period of about six years. Since this orbit was more typical of a comet than an asteroid, Carolyn studied the discovery images closely to see if our new asteroid looked fuzzy in any way. It did not; however, I was scheduled to observe with the much larger sixty-one-inch telescope and CCD system with my colleague Steve Larson of the Lunar and Planetary Lab. I suggested to Larson that our powerful setup might detect a faint coma of dust, the signature of a comet, surrounding the object.

When the first image appeared on our computer screen we saw just a field full of stars, as well as faint lines that showed the weak areas of the chip. Then we flat-fielded the image, a procedure calling for merging the image with a second image of a flat, lit surface (like the inside of the observatory dome). Combining the flat-field image with the original erases the defects in the CCD chip, leaving an evenly textured image with stars that appeared brighter and a background that appeared darker. And near the bottom of the field, one of those stars had a tail!

When we combined the images, we got seven views of the asteroid-turned-comet. The next day our message appeared in an IAU Circular titled "1990 UL3 = Periodic Comet Shoemaker-Levy 2 (1990p)." That also meant that our earlier Shoemaker-Levy comet, also periodic, would be now known as Shoemaker-Levy 1.

By February 1991, we had found two more Shoemaker-Levy comets. That month, Comet Levy 1990c, now much fainter than it was when it ruled the sky the previous summer, crossed the plane of the Earth's orbit and for two or three weeks sported a beautiful anti-tail. Composed of sunlit particles around the comet, the anti-tail points toward the Sun instead of away from it and becomes visible when the Earth crosses the plane of the comet's orbit. In May 1991, I observed my comet on the first anniversary of its discovery. Its show was over; it was very much faded and hardly visible at all.

TWO MINUTES

A week later, with dawn less than an hour away on the morning of June 10, I awoke to search the morning sky, but as is usual for that time of year, it was cloudy. Not wanting to give up just yet, I opened the sliding roof on my observatory shed and waited. Soon the first signs of dawn were appearing. I decided to search in the constellation of Aries, which was just coming up over the trees and seemed to be the only part of the sky that was clear. After one minute of slow sweeping, a bright fuzzy patch of light entered the field of view, and for a second a now-familiar "red alert" went off in my brain. But this fuzzy object was a distant galaxy called Messier 74. More than 200 years ago it duped French comet hunter Charles Messier, who sketched its position and then checked back later to see if it had moved, as all comets do. When the object still was frozen in the sky, he added it to his catalogue as he went on in search of slowly moving cometary prey. On this early summer morning in 1991, I did the same. As the sky continued to brighten, I moved Miranda over a few more fields. Another minute passed by, and then the mental alert went off again. There was another fuzzy spot.

For an instant I thought it was Messier 74 again. But this fuzzy was quite a bit brighter than the galaxy. With mounting tension I put in a higher power eyepiece and looked more closely. Where M74 had relatively sharp edges all around, this thing had a bright center, then faded off so slowly that I could hardly tell where it ended and the sky began. Sharp edges are characteristic of a galaxy filled with stars; the gradual fading of the edges is a comet's typical signature. This, I decided, was a comet. A closer look, and I noted a short tail pointing away from the Sun. I went inside to compose a message. Not a telegram this time, but e-mail, to Brian Marsden. An hour or two later, a one-page announcement circular appeared at observatories and universities around the world. This was a new periodic comet, it later turned out, that returns to the vicinity of the solar system every half century. Somehow, it had never been seen earlier, except possibly in 1499. In that distant year, Chinese and Korean observers observed a comet pass from Hercules through Draco and the Little and Big Dippers.[4] The orbit of that comet is so similar to that of Periodic Comet Levy that it could be the same comet. We will probably know the answer when it returns in the middle of the twenty-first century.

By the end of 1991, the Shoemaker-Levy team had discovered no fewer than eight new comets. Those comets that were found to have long orbits in which the comet would not return for at least two centuries were named Shoemaker-Levy, followed by a designation such as 1991d. Periodic comets were numbered. The last of these for 1991, Shoemaker-Levy 7, shed its light on our films one winter night during a ferocious wind storm. The wind was blowing so hard that the telescope's metal shutters were swinging about at the top of the telescope tube. To steady them Gene stood precariously on the top of an elevating chair, reached to the top of the telescope, grabbed the shutters, and held on tight. It worked: For the next few minutes the wind howled as Gene held on to the shutters with his hands and to the chair with his feet, trying to keep the scope steady and himself from tumbling off the chair. Meanwhile I struggled to keep the guide star centered as Carolyn asked whether this exposure was really necessary. "You never know what's in the field," Gene answered, "if you don't shoot it." He was right.

About forty-five minutes later we repeated the acrobatic exposure. When Carolyn scanned that infamous pair of films, she found the new comet that became known as Shoemaker-Levy 7.

In 1992 we found Comet Shoemaker-Levy 8, and a year later Shoemaker-Levy 9, the comet that collided with Jupiter. One more Shoemaker-Levy comet turned up in 1994, and on April 15 of that year, while searching visually, I discovered Comet Takamizawa-Levy, a comet then crossing the tiny constellation of Equuleus, the horse. As of today, our search goes on, both visually and with the Shoemaker-Levy Cometograph. It is an enjoyable pastime, for no matter how long we might search before our next comet is found, what we find on the road to a comet is what makes the search so rewarding.

STAR UPON STAR UPON STAR, OH MY!

The most recent examples of these "roadside discoveries" have to do with groups of stars that seem to be assembled in unusual shapes. Out in the observatory on the second night of the year 2000, Wendee and I took a series of photographs. While scanning one of the pairs of films, I found a ring of twelfth and thirteenth magnitude stars, open at its southern end. Brent Archinal, an expert on the clusters and associations of stars in the Milky Way told me that the ring had not been documented before, and suggested that I catalogue it as Levy-Wallach J2204.4+4509, according to its position in the constellation of Lacerta. We decided that the nickname of this interesting chain of stars would be Wendee's Ring. An ethereal sight on those discovery films, the ring is the first "discovery" I have ever made of something in the sky that will never move or change, at least on the timescale of a human lifetime.

On Christmas night, 2000, Wendee and I photographed an area that captures Equuleus, the little horse, the second smallest constellation in the sky after Crux, and possibly one of the least interesting in the sky. But not for me. Besides being the temporary home of a comet I found in 1994, it seems Equuleus contains a beautiful asterism consisting of about twenty to twenty-five faint stars that approximates the shape of a letter S. It's just north of the variable star RR Equulei. This "Equuleus S" asterism is cataloged as Levy-Wallach J2109.0+0618, Although both these groupings could be real strings of related stars, they are more likely chance groupings of unrelated stars. But they are part of this amazing sky through which we search. It is a sky that contains clouds of gas that mark the birth of new stars, stars that change in brightness, the fantastic swirls of the Milky Way, galaxies in an expanding universe—and through all of these gems, the stories of people who shared the wonder and made discoveries with it.

THREE

1572:
TYCHO BRAHE

A Swashbuckling Astronomer

Reach me down my Tycho Brahe—I would know him when we meet,
When I share my later science sitting humbly at his feet . . .

Though my soul may set in darkness, it will rise in perfect light;
I have loved the stars too truly to be fearful of the night.
        —Sarah (Sadie) Williams, 1869[1]

That famous star nail'd down in Cassiopee.
How was it hammer'd in your solid sky?
What pincers pulled it out again, that we
No longer see it, whither did it fly?
Astronomers say it was as least as high
As the eighth sphere. It gave no parallax,
No more than those light lamps that there we spy.
        —Henry More, 1647[2]

A large nobleman with a long red moustache and beard and the flair of a swashbuckler, Tycho Brahe walks onto this book's stage as our earliest example of an astronomical discoverer.

Some finds take years. Tycho's first two took seconds, and they had an indelible effect on the course of history. Born December 14, 1546, Tycho became one of the greatest astronomers who ever lived. He is credited with setting the observational groundwork that led to the confirmation that Earth is not in the center of the universe. Though he had the greatest observatory of his time on Earth, he preferred to be thought of as an aristocrat more than an astronomer. Allegedly he always wore full court regalia while observing because, most likely, that lent distinction to his pursuit. Science in the 1500s was looked down upon, and so were those who philosophized about it and studied it. Tycho did not want to let the fact that he practiced science interfere with his lofty reputation as a nobleman.

Tycho's family was very wealthy. His father owned a large tract of land in Scania, the southern part of what is now Sweden. His parents sent him to the university at Copenhagen to learn to be a statesman. His interest in astronomy dates from the August 21, 1560, eclipse of the Sun. The fourteen-year-old was fascinated by the ability of the astronomers of the time to predict so accurately the start and end of the solar eclipse.

By 1563 Tycho was becoming more serious about astronomy. A big and rare event was about to occur: a once-in-twenty-year "great conjunction" of Jupiter and Saturn in the constellation of Cancer. Great conjunctions occur every twenty years when these two mighty worlds appear close to each other in the sky. Jupiter takes twelve years to orbit the Sun once, and Saturn requires twenty-nine. Thus, every twenty years, as Jupiter seems to overtake Saturn, the two planets appear close to each other in a "great conjunction." (The most recent great conjunctions occurred around 1960, 1980, and 2000.)

Impressed by the predictions of the timing of the eclipse he had seen, Tycho expected the same accuracy in predicting the date of the closest conjunction of Jupiter and Saturn, and was puzzled when it didn't coincide with the prediction. His source, a book by Johannes Stadius, was based on the theory of Copernicus, and it missed the date of the conjunction by several days. Puzzled at this discrepancy, Tycho then checked the predictions for the date of this conjunction with the Alfonsine Tables, which used numbers according to Ptolemy's Earth-centered system. The result here was even worse. Realizing, correctly, that the problem must be poor observations, Tycho decided to build precise measuring instruments and collect accurate observations of the planets.

TYCHO'S NOSE

Three years after the great conjunction, Jupiter and Saturn were far apart in their positions in the sky and Tycho was a cocky student in the German city of Rostok.

There, the big man was partying at a pre-Christmas dance when he got into an argument with another party guest, Manderup Parsbjerg. The dispute was broken up, but at a Christmas party a few nights later they argued again, this time much more loudly. They decided to settle their dispute in a duel two nights later. Swords clashed. Although history doesn't confirm what happened to Parsbjerg, Tycho did not fare well. The tip of Parsbjerg's sword found its way into Tycho's nose, slicing it off at the bridge.

Tycho was able to stop the bleeding, but the resulting scar was unsightly and undignified. He took advantage of his own metal casting abilities to fashion a prosthesis of some skin-colored copper alloy that was attached to the bridge of his nose. Tycho turned what could have been a major embarrassment into a sort of mark of honor. This was not a man to be trifled with![3]

TYCHO'S STAR

Tycho did not learn to control his temper in the years after the duel. He lived for a time with his father, who died in 1571, and then with his uncle, where he developed his interests in chemistry and alchemy. Tycho was letting his astronomical interest lapse. This lull ended suddenly on a clear early October evening in 1572. He was stunned beyond words to see a bright new star adding to the five bright members of Cassiopeia. He did not believe the evidence of his own eyes, and asked his servants and local farmers to report what they saw as well. When he got over his surprise he began making detailed measurements of the position and brightness. His own measurements indicated that the supernova was in Cassiopeia at 0 hours, 22 minutes, and +63 degrees, 53 minutes.[4] The star has not been identified with certainty even today, so it is at least twenty-two magnitudes fainter than it was that night.

Tycho's measurements of this bright *stella nova* confirmed that it did not move relative to the distant stars, and therefore must belong to that outermost sphere, or area of space, that lay beyond the Moon. Visible in broad daylight on November 11, 1572, the star faded in brightness and disappeared in 1574. But the star also changed color as it faded. At the time, most scientists still accepted Ptolemy's Earth-centered universe theory, which also held that the eighth sphere, which contained the fixed stars, was supposed to be unchanging. By far Tycho's most important conclusion was that the star proved that the distant heavens are not, as had previously been thought, immutable.

URANIBORG

Shortly after the discovery of the new star, Tycho moved with his wife and growing family to Copenhagen. The place where he made most of his contributions to astronomy, however, was at his island observatory on Hven, some fifteen miles northeast of Copenhagen. Tycho was unquestionably lord over the entire island of Hven, an honor bestowed on him by none other than Denmark's King Frederick II, a man with a great interest in the sciences. (No longer a part of Denmark, Hven became a part of Sweden more than half a century after Tycho's death. Today Hven still remains a rural island, with a single village, Backviken, on its southeastern shore.)

On this island, Tycho lived in his home and observatory, called Uraniborg, the site of major observations from its completion in 1576 to the time he abandoned it in 1597. Never forgetting that he was a nobleman, he designed Uraniborg with the attributes of a castle. The main Uraniborg building was forty-nine feet square and some thirty-seven feet in height. Reflecting the cosmic vision of its builder, the building's four main doors faced the four cardinal compass points. Uraniborg offered some highly advanced features for a sixteenth-century establishment, including running tap water and central heating. The observatory included two features almost unheard of before or since. One was an author's dream of a paper mill and printing press, so that Tycho could publish his books and manuscripts independently, and the other was a prison.

OF QUADRANT AND ARMILLARY SPHERE

The "great quadrant" was one of several instruments at Uraniborg. It was made of metal, inscribed with a scale divided into small fractions of a degree, and included a sighting arm. Its radius was more than six feet. This big quadrant allowed Tycho to measure star declinations with greater accuracy than had been achieved before. (Declination is the same as earthly latitude projected onto the sky; if you are at latitude 45 degrees north, for example, a star at the same declination will be straight overhead when it is at its highest point in the sky.) Tycho also designed and built a large equatorial armillary sphere. It consisted of a big sphere, a rod, a sight with two narrow slits that slides along the sphere's circumference, and an arm that connects the sight to the center of the sphere.

To measure the position of a star or planet with this large instrument, Tycho sighted it with each of the two slits by moving the entire sphere around its central rod, and then raising or lowering the arm. He would then measure the angle of the arm, which would give him the object's declination, and then the angle of rotation of the entire circle to give the right ascension. (Right ascension is the projection of earthly longitude onto the sky.) With his instruments Tycho was able to measure the positions of Mercury, Venus, Mars, Jupiter, and Saturn, and almost eight hundred stars, over many years. With these measurements Johannes Kepler was able to form his three laws of planetary motions. Subsequently, Newton used Kepler's laws to formulate his theory of gravitation.

TYCHO'S COMET

The observatory was only a year old when the owner went outdoors fishing in one of the small ponds on the island. Tycho's ponds did double duty of being a repository of fish as well as powering his paper mill. The Sun had just set on that evening of November 13, 1577, and in the darkening sky he could make out a bright object in the southwest. He was as amazed as he had been five years earlier when he saw the star in Cassiopeia. This new object was as bright as the other had been! But this clearly was no star. As the sky continued to darken he could make out the long tail that stretched through Sagittarius and Capricornus. Tycho had independently discovered a comet.

Although Tycho never took credit for the comet's discovery— it was first seen by skywatchers in Peru on November 1, in London on the next night, and by seamen a week later—he did, however, sight it without any foreknowledge that it existed. Further, there is no question that he studied this comet with more care and thoroughness than anyone else. At the time, the prevailing view about comets was that of Aristotle, who held that they are exhalations from Earth's atmosphere. With his accurate instruments, Tycho set about that very night to measure the comet's changing position among the stars.

Tycho observed this comet at every opportunity from November 13 to the night it apparently faded beyond visibility on January 26, 1578. By comparing his observations with those of other astronomers throughout Europe, he was able to show that the comet appeared in the same position, relative to the background of stars, for any observer throughout Europe. Had the comet been in our own atmosphere, the positions from different cities would have differed. Thus, he concluded that the Comet of 1577 was farther from Earth than the Moon.

Tycho was not the first to suggest this, but he was the first to establish that fact *based on observation*. He did so eleven years after the comet was but a memory, in his 1588 book *De mundi aetherei recentioribus phenomenis* (Concerning the recent phenomena of the aethereal region).[5] Although the great comet was never formally named, it seems fitting to call it "Tycho's Comet" after the man whose relation with it began with an independent discovery, which he followed up with months of careful observation and years of analysis.

THE GREAT CONJUNCTION OF 1583

Six years after Tycho's Comet came the next great conjunction, in 1583. For the astrologers of the time, these conjunctions were crucial. The earlier one took place on the boundary between Cancer and Leo. Astrologers called Leo one of the signs of the "fiery trigon"; the other two signs of which were Aries and Sagittarius, so called by astrologers who divided the zodiacal constellations among Aristotle's four elements of earth, air, fire, and water. The great conjunction of 1583, astrologers noted, would be the last for centuries in any of the watery trigon constellations.

When Jupiter and Saturn got as close to each other as they would get during this conjunction, however, the Sun was in the way so that they could not be directly observed. Thus there was a controversy as to whether it took place in the watery trigon sign of Pisces, or in the adjoining trigon sign of Aries. Tycho took meticulous observations of Jupiter and Saturn, and confirmed that the conjunction had indeed taken place in Pisces. The following one, in 1603, was within the fiery trigon. So much was written about trigons that the term became part of the general vocabulary of the time. In fact, this was a time of not one but two great conjunctions, one between Jupiter and Saturn, the other between the sky and literature: Shakespeare referred to it in *Henry IV*, Part 2, where Prince Hal says, "Saturn and Venus this year in conjunction? What says th' almanac to that?" The prince's humorist Edward Poins replies, "And look whether the fiery Trigon, his man, be not lisping to his master's old tables, his note-book, his counsel-keeper."[6] Written around 1598, Shakespeare was possibly harking back to the controversy of the great conjunction of 1583.[7]

Tycho's contribution to astronomy might have exceeded matters purely astronomical. Owen Gingerich, an astronomical historian at Harvard, points out that Tycho's observatory at Uraniborg was not far from Hamlet's castle in Elsinore; on a clear day Tycho could look out across the water and see it. One of Tycho's books, printed at his observatory in 1596 and titled *Epistolae*, boasted a portrait of Tycho surrounded by the coats-of-arms of his ancestors, which included Rosenkrans and Guldensteren. When Shakespeare published *Hamlet* in 1603, he chose those names for Hamlet's university colleagues. By what means did Shakespeare obtain a copy of Tycho's somewhat obscure book? He might have seen it or received it from Thomas Digges, a famous astronomer in London who knew Tycho and who lived near Shakespeare; possibly Shakespeare enjoyed the unusual names enough to make use of them.[8]

Within a year of the great conjunction, Tycho built Stjerneborg, or Star Castle, Observatory, an annex to Uraniborg. Surrounded by a stone fence some fifty-seven feet square, Stjerneborg's rooms were partly underground to provide protection from wind and weather. They housed Tycho's assistants and his visiting students. Often they would make observations independently, comparing them later to those made at the main facility at Uraniborg. The beautiful domed buildings contained underground, heated rooms.[9]

TYCHO'S COMPROMISE

Tycho was a product of an age that began when Nicholas Copernicus died on May 24, 1543, receiving on that day the first copy of his newly published book *De Revolutionibus*. At that moment he left the most important scientific work since the height of Greek civilization. *De Revolutionibus* was a sensation among the intellectuals of the time. Incidentally, the first political use of the word "revolution" appeared in the year 1600, to mean "a complete overthrow of the established government in any country or state by those who were previously subject to it."[10]

Copernicus's theory proposed that the Sun, and not the Earth, was in the center of the universe, and it sparked a headlong rush for observations to prove or disprove it. Tycho, with his beautiful instruments of measurement, was glad to test it. He accomplished for Renaissance observing what the Greek astronomer Hipparchus did for Greek observing. Tycho's ideas were based on meticulous observation, rather than the theoretical arguments that led philosophers to accept the *Almagest* of Ptolemy, a work that championed the Earth-centered universe and which was considered as virtual gospel since its appearance around 150 C.E. Far in advance of his time, Tycho's idea of basing arguments on observation would not gain support until 1620, when Francis Bacon published *Novum Organum* and invented what we now call the "Scientific Method."

Tycho shared Copernicus's view that the retrograde motion of the outer planets could not be explained if these planets orbit the Earth. (Retrograde motion takes place when Mars, Jupiter, and Saturn, or any world in our solar system farther out from the Sun than Earth, appear to stop their eastward motion among the stars, then back up, and head westward for a few months before slowing down and resuming their eastward motion. The effect is actually similar to passing a slower car on a highway; as you overtake it, the car appears to move backwards.) However, despite Tycho's scientific inclination and love of observing, he was opposed to the new theory of Copernicus. "His objections do not appear to have arisen through religious bigotry," wrote British astronomer Fred Hoyle four hundred years later.

> Rather, it seems, they sprang from a characteristic that marks almost all great observers: that the world as they see it has a more immediate emotional reality than it does for an ordinary person, and very much more than it does for the theoretician.

This mystical relationship between observer and object seems to arise only as a result of light actually entering the telescope or the eye, and does not exist, for instance, when one is looking at a photograph of a celestial object. This psychological trait leads the observer to doubt the reality of situations in which he cannot establish the same physical contact.[11]

Tycho's observations said one thing, his heart another. He set to work on an arrangement where the planets could revolve about both the Sun and the Earth at the same time. While building evidence for his model, he determined the real cause of why predictions of planetary conjunctions were still off. The theory of Copernicus was correct, he determined, but the accuracy of the planetary positions that had been obtained with the older telescopes was poor. With his new devices, Tycho's measurements were far more accurate. Tycho's result was a compromise between Ptolemy's Earth-centered universe and the heliocentric model devised by Copernicus.

In 1583, Tycho proposed that all five planets then known—Mercury, Venus, Mars, Jupiter, and Saturn—orbited the Sun, which in turn orbited the Earth. Unquestionably Tycho's model was a step backward, a last-ditch effort to keep the Earth at the center. It was a beautiful model worthy of a philosopher, but unworthy of the great observer that Tycho was. "But what an untoward broken system of the world this of Tycho's is," wrote poet Henry More in 1647, "in comparison with that of Copernicus will appear even at first sight, if we do but look upon them both."[12]

LAST YEARS

In his final years Tycho had sparked so many arguments with so many people that he worried that his precious equipment might be confiscated. As a feudal lord over his island, Tycho was fierce and oppressive to the local farmers, who finally rebelled. Meanwhile, he was rapidly falling out of royal favor. The new king did not share his predecessor's interest in science, but he did hear the complaints of the peasants on Tycho's island. With the loss of his pension, Tycho left Hven with his family, first for Copenhagen, and then to wander around Germany for two years. Finally, in 1599 Rudolph II of Germany allowed the Brahe family to set up residence in a castle about twenty miles from Prague.[13] He had his twenty-eight instruments brought from Uraniborg and reassembled at his new castle. He began his observing again, adding this time an assistant, Johannes Kepler. A devoted assistant and friend, Kepler seemed to have infinite patience for the observing master, and in the two years that they worked together, he developed an appreciation for the accuracy of the observations that Tycho made.

During the summer of 1601, Kepler noticed that "the feebleness of old age" had struck his master.[14]

Tycho Brahe died at age fifty-five. On October 13, 1601, while at supper, he suddenly became ill. "On returning home," wrote his biographer J. L. E. Dreyer, "he suffered greatly for five days. During the night before the 24th October he was frequently heard to exclaim that he should not appear to have lived in vain *ne frustra vixesse videar*."[15] Tycho's unusual life is reflected in the fact that the Moon's most dramatic crater was later named after him. Created by the impact of a comet or asteroid some 100 million years ago, crater Tycho is surrounded by bright rays of debris scattered halfway around the Moon.

Even on Earth, Tycho would not be allowed to rest completely in peace. Three hundred years after his birth, the City of Prague permitted two scientists at the local university medical school to open his ornate tomb. Inside they found two skeletons. One was of a woman—his wife, most likely—with hands folded; the other was of a large man, still with evidence of moustache and beard, covered in a red robe. The upper end of the man's nasal opening had a green stain of copper that was probably left over from a coppery object that had been in contact with the bone for a long time. So the story of Tycho's artificial nose seems to have been confirmed.[16]

One other part of Tycho's life did not end with his death. As Tycho lay dying, he begged Kepler to pay attention to his system of the universe, where all the planets go around the Sun while the Sun revolves around the Earth. Although he knew the Tychonic system was wrong, Kepler promised that he would not forget it, and though his results clearly supported the Copernican theory, he often referred to Tycho's model in his later writings. Kepler brought his own idiosyncrasies to the world of science; he crafted some of the best physical laws science has ever seen, but he used these laws to produce horoscopes for friends and associates! Kepler's life was as troubled as Tycho's in some regards; he had to defend his mother when she was accused of witchcraft, and his own marriage was not a perfect one. Also, for someone with the greatest respect for Tycho's observational ability, Kepler did not enjoy observing.

Had Tycho lived but another three years, he would have joyfully observed a second *stella nova*. Instead, Kepler took full advantage of this eruption of a supernova in Ophiuchus. Not quite as brilliant as Tycho's star in Cassiopeia, Kepler's star became as bright as Jupiter. There was a second difference. While Tycho's star alerted one man, Tycho himself, to the idea that the heavens are not unchanging, Kepler's star arrived to a generation far more attuned to the changing ideas about the universe. Kepler's star could not have been better timed. It burst into the sky only a few months before a series of eclipses enriched the debate all over Europe as to whether the stars and planets exert an influence on humanity. While Kepler was having fun with horoscopes, Shakespeare was insisting, through Cassius in *Julius Caesar*, that

The fault, dear Brutus, in not in our stars,
But in ourselves, that we are underlings.[17]

Thanks to great dramatists like Shakespeare, the public was increasingly aware by 1604 that things were changing in the heavens. What it didn't know yet was that yet another stargazer was on the scene who would, within five years, change humanity's perceptions of itself and the cosmos forever.

FOUR

1610:
GALILEO AND THE
INTERPRETATION
OF DISCOVERY

A man is here revealed who possesses the passionate will, the intelligence, and the courage to stand up as the representative of rational thinking against the host of those who, relying on the ignorance of people and the indolence of teachers in priest's and scholar's garb, maintain and defend their positions of authority. His unusual literary gift enables him to address the educated men of his age in such clear and impressive language as to overcome the anthropocentric and mythical thinking of his contemporaries and to lead them back to an objective and causal attitude toward the cosmos, an attitude which had become lost to humanity with the decline of Greek culture.
—Albert Einstein, July 1952[1]

Imagine somehow being able to trade places in time, so that for a period of ten minutes, you could be a part of the early seventeenth century, and that in your place, a person from that period could be a part of life in the 2000s. Imagine how interesting such a switch would be! I would love to bring Galileo into the near present, and see his expression on being told that the planet that he studied with such enthusiasm was visited at the end of the millennium by a space carriage of sorts, named *Galileo* in his honor. What would he think when told that "his" planet Jupiter suffered a collision with a comet, and that this collision was also observed by that same spacecraft? I wonder what questions he would have about these amazing events.

Among the many scenes in Galileo's life that I would like to see, two stand out. The meeting between Galileo and John Milton in 1638, discussed previously, is certainly one of those. The other took place much earlier in Galileo's life, at the table of the Grand Duke of Tuscany in September 1611. This dinner took place the year following Galileo's discoveries of the moons of Jupiter, the phases of Venus, and the spots on the Sun. Besides the Grand Duke and Galileo, two cardinals dined there that evening. They spoke of the nature of ice, and why objects floated in water.

One of the two cardinals argued against Galileo during the dinner, restating instead the position of Aristotle. At dinner that evening, both sides of the argument were debated. One was based on Aristotle's position that bodies seek their natural places (fire and air are high; earth and water low) by some magical property they possess.[2] The other cardinal argued in favor of Galileo's position, supporting both the idea and the fact that Galileo had come to it via the route of observation and experiment. That friendly and progressive cardinal was none other than Maffeo Barberini, the same man who would soon be elected to the pontificate as Pope Urban VIII, and who would later spearhead the downfall of Galileo.

EARLY LIFE

Although the road to scientific discovery, knowledge, and understanding is fraught with many bumps, I do not know of a case where the consequence of a specific discovery was more traumatic than that of Galileo. Born in Pisa on February 15, 1564 (Julian Calendar or Old Style), Galileo shares his year of birth with William Shakespeare. His first name, Galileo, derives from an interesting Tuscan custom of the time that the first name of the firstborn child would be a repeat of the surname.

Even though Galileo lived to age seventy-eight, extraordinarily long for his time, he was often ill with arthritis or rheumatism. This condition might have been linked to a near-fatal circumstance that he suffered in his youth. There are two versions of what happened that hot summer of 1593. In one, Galileo and a group of friends, while walking on a very hot summer day, found a cave where cool air offered relief. Taking their clothes off, they lay down to rest and nap, unaware that the air within the cave was toxic. Some of the young men died from the poison air, and virtually all of them became sick.[3] A different version suggests that after their long hike, they ended up at a lawyer's summer villa. After a evening of drinking and conversation, Galileo removed all his clothes and lay down to sleep in a portion of the house cooled by a draft from a nearby spring. He caught a bad cold, which was later complicated by the onset of the chronic arthritis which plagued him for the rest of his life.[4]

The road Galileo took during his life almost turned away from science completely when he went to Vallombroso as a youth to be educated by monks. It was probably during this time that Galileo witnessed a hailstorm. As pieces of ice noisily struck the ground, Galileo noticed that the larger, heavier hailstones did not fall faster than the smaller ones. Noting his son's promising mind, Galileo's father, Vincenzio Galilei, a composer, eventually removed his son from Vallombroso. In 1581 Galileo enrolled at the University of Pisa's school of medicine. Around this period he also worked out the mathematics of the pendulum as a device for telling time. His discovery that no matter how forceful the push of the pendulum, the cycle of sway took the same amount of time, occurred as he was sitting in the cathedral at Pisa in 1581. He used his own pulse as a chronometer.

Two years later he met Ostilio Ricci, who introduced Galileo to the practical uses of mathematics. In 1589 Galileo was hired as chairman of the mathematics department at the University of Pisa. His first treatise, *On Motion*, was completed there. It was this interest that led to Galileo's famous demonstration on the Leaning Tower of Pisa during which he dropped objects of two different weights from the tower. He didn't drop a rock and a feather, as some have suggested, but two different-sized samples of the same substance. Both objects hit the ground at the same time in this successful demonstration that contradicted the belief of Aristotle. Did this event, in fact, actually happen? Its only source is Vincenzio Viviani, the student who lived with Galileo when he was, decades later, under house arrest. Viviani likely heard the story from Galileo himself, so we are inclined to believe it even though the two objects Galileo dropped should not have landed exactly simultaneously, since the different weights were in fact affected differently by the air rushing around them. In a vacuum, however, they would have hit the ground at the same time. A second difficulty with this is that not a single existing letter or piece of writing has mentioned the demonstration, even though, according to Viviani, the whole student population of the university was present to watch.[5]

PADUA AND THE TELESCOPE

In 1592, Galileo became chair of mathematics at the University of Padua, where he remained for the happiest eighteen years of his life. While there he lived with Marina Gamba, a Venetian woman with whom he had two daughters and a son. The elder daughter, Maria Celeste, always had a close relationship with her father. Born in 1600, she was originally named Virginia after Galileo's sister. (At the age of thirteen, she was placed by Galileo in the Convent of San Matteo, in Arcetri, where she changed her name. Galileo put her there supposedly for her own protection, because he believed that since the girl had been born out of wedlock she could never marry. This harsh treatment on Galileo's part really sentenced her to a life of poverty, and yet she always remained close to her father).[6]

One year after becoming chairman, Galileo invented the thermometer, a device that took advantage of his experiments that showed that the density of a liquid changes with temperature. His first one involved a small glass sphere at the end of a glass stem twenty-two inches long in a beaker of water. When he held the sphere with his hand, the water rose inside the stem. Later he refined the experiment using wine instead of water. He placed several glass spheres, each containing a liquid, into another liquid. He found that as the temperature fell, the spheres rose. Galileo's own experiments had confirmed the result of Archimedes, who, sometime before 212 B.C.E. first discovered the idea that bodies are buoyed up in proportion of the amount of water they displace. The story is told that when Archimedes discovered that principle, he jumped out of his bath and ran naked outside his home, shouting "Eureka!" When the Shoemakers and I discovered an asteroid in an orbit that follows closely that of Mars, we named it Eureka to honor the spirit of exhilaration that accompanies a discovery. Flush with the discoveries of 1610, Galileo must have shared Archimedes' feeling that night.

By 1602 Galileo was heavily involved in mechanical inventions. One was his thermoscope, yet another kind of thermometer in which he fashioned a glass bulb with a thin neck and then held it upside down in water. The height of the water in the neck indicated the approximate temperature of the air around it.

The appearance of Kepler's Star, the supernova of 1604, got Galileo's full attention; he frequently discussed the significance of this star in public lectures. By 1609, Galileo was in the midst of completing his work on mechanics and strength of materials, and preparing it for publication. That work was interrupted when he heard about the extraordinary invention of a spyglass by the Dutch lens grinder Hans Lippershey. Consisting of two convex lenses separated by a specific distance, the spyglass was able to magnify distant scenes.

Lippershey is not the first to have invented the telescope. Roger Bacon probably assembled one earlier and the Englishman Leonard Digges, who died in 1571, possibly assembled one.[7] However, whether either actually put their telescopes to use looking at the heavens is unknown. In any event, these earlier attempts at telescopes did not attract anywhere near the interest that Lippershey's invention did, and Lippershey was successful only because his invention came to the attention of Galileo.

If we don't know when, and by whom, the telescope first appeared, we also do not know who first used one to look at the stars. Was it the English mathematician Thomas Harriot or the German scientist Simon Marius or was it Galileo? Possibly both of these other men pointed their telescopes skyward, but if they did, and if they performed the extremely important additional step of recording what they saw, these records have apparently not survived. Galileo assembled a more powerful telescope, used it to study the sky, took copious notes and made drawings, and announced his results in writings that most of the literate population could understand.

We do know that 1608 was a crucial year for Holland, a country fighting for independence from Spain's Philip II. Prince Maurice, head of the Belgian army and a man interested in science, would have made good military use of Lippershey's invention. We also know that Hans Lippershey submitted a patent application for his spyglass to the States-General, the governing body of the Netherlands, on October 2, 1608.[8] By the early summer of 1609, Fra Paolo Sarpi learned of this extraordinary device and told his friend Galileo of it. Visiting Venice at the time, Galileo hurried back to Padua. He quickly made a telescope that magnified objects by nine times. He soon discovered that the Moon's surface was filled with mountains and craters, and he was impressed that it looked somewhat like that of the Earth.

THE MOONS OF JUPITER

By January 1610 Galileo had made a thirty-power telescope. On the evening of January 7, he took advantage of the clear sky and pointed the telescope toward Jupiter. The telescope's narrow field of view made it difficult to find Jupiter, but when he did, he saw a retinue of three stars, one to the right of Jupiter and two more on the planet's left: What happened next deserves to be told in Galileo's own words:

> On the seventh day of January in this present year 1610, at the first hour of night, when I was viewing the heavenly bodies with a spyglass, Jupiter presented itself to me; and because I had prepared a very excellent instrument for myself, I perceived (as I had not before, because of the weakness of my previous instrument) that beside the planet there were three starlets, small indeed, but very bright. Though I believed them to be among the host of fixed stars, they aroused my curiosity somewhat by appearing to lie in an exact straight line parallel to the ecliptic, and by their being more splendid than others of their size. Their arrangement with respect to Jupiter and each other was the following:

> *East*   *   *   **O**   *   *West*

> that is, there were two stars on the eastern side and one to the west. The most easterly star and the western one appeared larger [probably brighter] than the other. I paid no attention to the distances between them and Jupiter, for at the outset I thought them to be fixed stars, as I have said. But returning to the same investigation on January eighth— led by what, I do not know— I found a very

different arrangement. The three starlets were all to the west of Jupiter, closer together, and at equal intervals from one another as shown in the following sketch:

East   **O**   *   *   *   *West*

At this time, though I did not yet turn my attention to the way the stars had come together, I began to concern myself with the question how Jupiter could be east of all these stars when on the previous day it had been west of two of them. I commenced to wonder whether Jupiter was not moving eastward at that time, contrary to the computations of the astronomers, and had got in front of them by that motion. Hence it was with great interest that I awaited the next night. But I was disappointed in my hopes, for the sky was then covered with clouds everywhere.

On the tenth of January, however, the stars appeared in this position with respect to Jupiter:

East   *   *   **O**   *West*

that is, there were but two of them, both easterly, the third (as I supposed) being hidden behind Jupiter. As at first, they were in the same straight line with Jupiter and were arranged precisely in the line of the zodiac. Noticing this, and knowing that there was no way in which such alterations could be attributed to Jupiter's motion, yet being certain that these were still the same stars I had observed (in fact no other was to be found along the line of the zodiac for a long way on either side of Jupiter), my perplexity was now transformed into amazement. I was sure that then apparent changes belonged not to Jupiter but to the observed stars, and I resolved to pursue this investigation with greater care and attention.

And thus, on the eleventh of January, I saw the following disposition:

East   *   *   **O**   *West*

There were two stars, both to the east, the central one being three times as far from Jupiter as from the one farther east. The latter star was very nearly double the sized of the former, whereas on the night before they had appeared approximately equal.

I had now decided beyond all question that there existed in the heavens three stars wandering about Jupiter as do Venus and Mercury about the sun, and this became plainer than daylight from observations on similar occasions which followed. Nor were there just three such stars; four wanderers complete their revolutions about Jupiter, and of their alterations as observed more precisely later on we shall give a description here. Also I measured the distances between them by means of the spyglass. . . . Moreover I recorded the times of the observations, especially when more than one was made during the same night— for the revolutions of these planets are so speedily completed that it is usually possible to take even their hourly variations.[9]

In these historic words, Galileo recorded the presence of four moons orbiting Jupiter, a solar system in miniature. His telescope had found something entirely new, something whose implications were stunning.

FLORENCE

Galileo also used his telescope by day to study the Sun, and he carefully recorded the daily positions of the sunspots he saw. Without any filtering system, the dazzling Sun, even through his small telescope, might have led to his later blindness. Naming Jupiter's moons "the Medicean stars" in honor of the Duke of Tuscany, he hoped to be appointed the Duke's Chief Mathematician and Philosopher. Suitably impressed, the duke did indeed offer Galileo that appointment, and later in 1610, Galileo resigned his position in Padua and moved to Florence. That fall, when Venus finally came out of its conjunction with the Sun, Galileo eagerly turned his telescope to it. He was uncertain what he would see. Venus was capable of being much brighter than Jupiter, and perhaps it had a family of moons as well. Over the following few months he discovered no moons, but that Venus showed phases like the Moon.

Galileo's discoveries were startling. For the first time in almost 2,000 years, find after find was painting a picture of a totally new kind of solar system. Galileo even noted something unusual about the faint light, or Earthshine, that the dark side of the Moon shows when it is at crescent phase: the farther from the lit edge of the Moon an observer looks, the brighter the Earthshine is. Galileo added that this "secondary light" is more prominent when the Moon is close to the Sun, as in its crescent phase, and fainter as the phase edges past the quarter. Through his own observations, Galileo concluded that Kepler was right in suggesting that the Earth's own reflected light is the cause of Earthshine. When the Moon is a crescent as seen from Earth, the Earth would appear close to full as viewed from the Moon, and its light would be consequently greater. The Earth's phase, in other words, complements the Moon's phase.

Galileo noticed something that virtually everyone takes for granted. That Earthshine is much brighter when the Moon is at a thin phase can be confirmed, as Galileo himself suggests, by blocking the bright part of the Moon with a roof or chimney and viewing the faintly lit portion against a dark sky.[10]

Like any totally new argument, Galileo's observations and conclusions were accepted by some, snickered at by others, and detested by Galileo's most vicious enemies. By 1611, opposition was mounting; Nicolò Lorini, a Dominican, officially notified the Roman Inquisition that Galileo's views should be investigated as possibly heretical. Galileo rewrote his explanations and offered them to friendly church officials.

The astronomical discoveries were not the only problems confronting Galileo by 1615. His earlier research on the densities of bodies in water was being attacked as well. Galileo cleverly asked his friend Benedetto Castelli, then chairman of the University of Pisa Department of Mathematics, to handle the problem concerning objects in water while he concentrated on the far more serious accusations against him that grew out of the discoveries made with his telescope. During much of 1615 Galileo was ill, but by the end of the year he traveled to Rome to try to clear his name.

Galileo's public debates on the issue probably did more harm that good. Pope Paul V appointed a commission to study the Copernican theory, and its report was harsh. By order of the pope, Cardinal Bellarmine told Galileo in March 1616 that he must abandon his belief that the Earth travels around the Sun, since, in the opinion of the commission and the pope, that view was contrary to Scripture. If Galileo continued to express these views, the pope added, the Holy Office would issue a formal order forbidding his teaching of Copernican views on pain of imprisonment.

Galileo returned to Florence in the middle of 1616, the year of Shakespeare's death. Had he passed away that year at age fifty-two, he would have died at the pinnacle of his career. But Galileo lived on, sometimes in severe illness and discomfort, for another twenty-six years. In 1616, the Roman Inquisition took its first step against him.

FIVE

1632:
GALILEO AND THE
CONSEQUENCE
OF DISCOVERY

Et ei dicto quod dicat veritatem, alias devenietur ad torturam.
—official minutes of Galileo's trial, 1632,
stating that Galileo was under a threat of torture.[1]

We condemn you to the formal prison of this Holy Office during our plea-
sure and by way of salutary penance we enjoin that for three years to come
you repeat once a week the seven penitential Psalms.
—From Galileo's sentence, 1632[2]

Saddened by the 1616 injunction from Pope Paul, but never giving up, Galileo decided to put his Jupiter discovery to a different purpose. Could sailors use the positions of the satellites of Jupiter to determine a ship's longitude? The idea was sound, and the problem Galileo addressed was crucial, for at the time ships had no reliable way of determining their longitude. Latitude was a simple problem; a single nighttime observation of the altitude of a star above the horizon as it crossed the meridian, or its highest point in the sky, would yield the latitude of a ship. But longitude was far more complicated; calculating a ship's longitude depended on a way of measuring time at sea, which the clocks of that era could not do. Galileo hoped that since the positions of Jupiter's satellites could be predicted with a great degree of accuracy, the planet could act as a sort of cosmic clock whenever it was above the horizon. Sailors would plot the positions of Jupiter's satellites and compare them to published tables that showed where the satellites were during the course of a day. Galileo's plan was brilliant, but in those days telescopes were very difficult to use even on land, let alone from a moving ship.

THE COMET QUARTET OF 1618–1619

Except for a single comet in 1615, no major comet had appeared after Galileo's first use of the telescope. The comet drought was broken in the late summer of 1618, when the first of a procession of four bright comets arrived. Kepler, with other observers, found the first one, a splendid morning object in the constellation of Leo with a long tail pointing to the west. "In the month of August," wrote Jacobus Mascardus, a Jesuit priest, "news was brought to us from many parts of Italy that during that period a comet was seen licking the hind feet of the Great Bear." Kepler had a telescope by this time and he made what is probably the first telescopic observation of a comet. The comet was fading by the middle of September.[3]

By the middle of November a second comet punctuated the southwestern sky. Its head was in Libra, and its tail stretched for forty degrees in length (eighty full-Moon diameters!). A third comet emerged at the same time, creeping across the sky from Libra toward the west. The final member of the foursome was sighted on February 14, 1619, in the southeast, with a tail covering more than half the predawn sky.

The excitement generated by the quartet's arrival ignited a debate on the nature of comets between Galileo and several Jesuit astronomers. One astronomer, Horatio Grassi, suggested that comets have circular orbits around the Sun.

His argument was beautifully phrased: "I have believed that the comet," he wrote, "shining on all directly from the same place and appearing the same from all sides, must be considered as worthy of the heavens and very near to the stars." With a flourish Grassi concluded with a quote from Horace: "With my head exalted I shall touch the stars."[4]

Another Jesuit astronomer, Mario Guiducci, wrote a *Discourse on the Comets*. In it he wrote of comets behaving "like Penelope unraveling the cloth with one hand as fast as she weaves it with another." Galileo decided to begin writing again, possibly because he believed that the Jesuits were taking undue advantage of the comets to support the Tychonic theory of the universe. In 1623, he published *Il Saggiatore* (*The Assayer*) in the form of a letter.

The Assayer was a mostly theoretical document about comets. Unlike his earlier writings about the solar system, it is not an observational book and does not include any mention that he observed any of these comets through his telescope. During the fall of 1618, however, Galileo was bedridden with arthritis and a hernia, and was quite unable to use his telescope. The book does argue about the orbits of comets. The orbits could not be circular, Galileo argued against Grassi, since it is clear that all comets do not return again and again. Comets were formed instead, Galileo countered, by sunlight reflecting from thin vapors in Earth's upper atmosphere. Galileo's strange acceptance of Aristotle's theory of comets would seem to put him on the wrong side of history. Stillman Drake, one of the most respected present-day Galileo scholars, suggests that Galileo was fighting a different battle here. Galileo was not really arguing about the nature of comets; he was trying to debunk the Tychonic system. Besides, Drake continues, Galileo was right that comets do not shine by their own light.[5]

The Assayer was a pivotal volume for two reasons. First, it was so popular that Galileo was encouraged to proceed with his *Dialogue of the Two World Systems*, the work that led directly to his final downfall. Second, thanks to this book, the rift between Galileo and the Jesuit astronomers widened into a chasm that could never heal. In its opening pages, Galileo wrote angrily that "some, conquered and convinced by my reasons, attempted to rob me of that glory which was mine by pretending not to have seen my writings and subsequently trying to make themselves the original discoverers of such impressive marvels."[6] Sadly, the one person who had written about sunspots was Christopher Scheiner, and he thought that besides himself and Galileo, no one had written about sunspots. Scheiner was enraged with Galileo at what he considered was a personal attack.

This was truly unfortunate. Scheiner and Galileo had been close friends, but their rivalry over who discovered the sunspots first embittered them both. *The Assayer* made things worse. When Galileo published his *Dialogue,* he knew that Scheiner was planning to publish *Rosa Ursina,* a text whose purpose seemed to be to extract vengeance on his old friend for the sunspots, and for his attack on Grassi in *The Assayer*. The worst part of Scheiner's attack is that deep down, this astronomer was a Copernican. Apparently his anger at Galileo was so great, or perhaps his desire to make points with his superiors was so great, that he buried his own scientific beliefs.[7] With the publication of *The Assayer*, another turn in the road to Galileo's fall was irrevocably taken.

CHANGE IN FORTUNE

Pope Paul V died in 1621, and Pope Gregory XV, who ruled but briefly, died in 1623. Imagine Galileo's excitement when his old friend Maffeo Barberini, the same man who argued convincingly in favor of Galileo, was elected to the papacy. The new pope couldn't wait to see his scientific friend. "Is Galileo coming," he asked with the impatience of a child, "When is he coming?"[8] Galileo planned a visit to the Holy Office to present his friend with his latest device, a microscope. They took long walks in the gardens and discussed all manner of scientific questions. Although Galileo understood that he was no longer talking with Barberini but with Pope Urban VIII, he felt assured by the pope that his enemies would never get the better of him.

Galileo thought that the time to publish had come. Nine years after the election of the new pope, in 1632 Galileo finally had permission to publish his *Dialogue on the Great World Systems*. The book was an instant hit throughout Europe, but it enraged the pope. "Your Galileo," the pontiff barked to one of Galileo's friends, "has ventured to meddle with things that he ought not and with the most grave and dangerous subjects that can be stirred up in these days."[9] The book was a catastrophe for Galileo. Somehow Galileo had misread one of the pope's objections. When they were walking in the pope's garden, and Galileo was explaining his view of the tides, Urban replied that God could create and manage events like orbits and tides any way he wished.

In the *Dialogue* Galileo, through his character Salviati, accused those who maintain that the Earth is the center of the solar system of reasoning preposterously. "Over the long run my observations have convinced me that some men, reasoning preposterously, first fix some conclusion in their minds which, either because it is their own or because of their having received it from some person who has their entire confidence, impresses them so deeply that one finds it impossible even to get it out of their heads."[10] That was the way of the old Earth-centered system of Ptolemy, concludes Stillman Drake. "I think it sad," he adds, "that the Inquisitors favored scholars rather than men of knowledge and good sense."[11]

That, however, seems precisely how Galileo's troubles began. They mounted later when things became both political and personal. When the pope finished reading the *Dialogue*, he felt that Simplicio, whose name Galileo actually took from a scholar from the sixth century, might be a characterization of him. The name certainly conjures up, however, the idea that Galileo intended this character to be simple-minded.[12] Some aides, particularly Scheiner, encouraged the pope in that belief.

Thanks to a campaign led by Scheiner and others, Urban VIII was convinced that Galileo had duped him into granting permission to publish a book that made him look foolish. The Simplicio aspect was a good part of the problem: "He did not fear to make sport of me," said the incensed pontiff.[13] That wasn't all: The pope decided that he had been fooled not just by Galileo but by his some of his own people. Now he was being told that the *Dialogue* was a greater threat to the Church even than the rebellion of Martin Luther. "The Pope lives in fear of poison," one observer wrote about the pope's feeling that his papacy was going nowhere. The *Dialogue* was just one more thing that happened at the wrong time. "At the hour of what he knows to be scientific victory," writes science historian Giorgio de Santillana, "Galileo is generously throwing the mantle of his intellectual prestige around the authorities so as to cover their past incompetence and obduracy."[14]

On October 1, 1632, there was a knock at Galileo's front door. It was the Inquisitor of Florence, with a summons for Galileo to appear in Rome within thirty days. In response, Galileo begged for a delay, claiming that his ill health meant that the journey by sea, during inclement winter weather, could well be fatal for him. The pope refused even to consider Galileo's request. He expected the summons to be carried out, and to be informed when Galileo was in Rome.

Galileo arrived shaken but ready for his trial. His months in Rome were mostly spent waiting—waiting for the trial to begin, waiting between sessions and waiting as negotiations were carried out. That Galileo was in acute pain, both emotional and physical, cannot be overstated. He suffered from indigestion, insomnia, the onset of blindness, and arthritis. His legs were so painful that at night he would scream in agony, and yet he had to beg for permission merely to walk outdoors to exercise to relieve the pain.[15]

During a good part of his life, and especially at this terrible time, Galileo kept in contact with his daughter, now named Maria Celeste. Their correspondence indicates that they were very close. "Most beloved Lord Father," she wrote on April 20, 1633, "alas you are detained in the chambers of the Holy Office; on the one hand, this gives me great distress, convinced as I am that you find yourself with scant peace of mind, and perhaps also deprived of all bodily comforts: on the other hand . . . I console myself and cling to the expectation of a happy and prosperous triumph. . . ."[16]

Finally, Vincenzio de Firenzuela, one of the Inquisitors, visited Galileo. Explaining that it was the will of the highest office of the Church, implying Urban VIII, that an example be made of Galileo, Firenzuela proposed a compromise in which Galileo asked for forgiveness for some misdeed, in exchange for a relatively light sentence. He would not have to recant his belief in the Copernican theory. Galileo wrote a long letter in which he blamed his book on what might charitably be called an excess of enthusiasm.

Fully expecting the proceedings to be ended with a mild sentence, Galileo waited anew as the tribunal deliberated. But something went wrong. According to Drake's speculation, the inquisitors might have worried that once Galileo was back at his villa, he would be free to speak openly about how he was brought to Rome because of a trumped up interpretation of the Codex of 1616. It could be a major embarrassment to the Holy Office. The final sentence, by decree of Urban VIII to be handed down in a formal ceremony complete with public apology, presented to Galileo, an old man dressed in a penitential white robe, was life in prison.[17] The only specific reference to an observation through Galileo's telescope was to his sunspot observations: He was accused of having printed "certain letters," entitled "On the Sunspots," in which he affirmed beliefs that the Church found heretical.

Galileo, feeling betrayed, was angry and stunned. "In the Galileo trial," wrote de Santillana, "the Inquisition was suborned into a command performance by an unscrupulous group of power politicians; even so, the behavior of the judges compares most favorably with that of . . . Henry VII's judges in a treason trial."[18] "My dearest lord father," daughter Maria Celeste wrote on July 2, 1633,

> now is the time to avail yourself more than ever of that prudence which the Lord God has granted you, bearing these blows with that strength of spirit which your religion, your profession, and your age require. And since you, by virtue of your vast experience, can lay claim to full cognizance of the fallacy and instability of everything in this miserable world, you must not make too much of these storms, but rather take hope that they will soon subside and transform themselves from troubles into as many satisfactions.[19]

Trying to take his daughter's advice, Galileo begged for a more lenient sentence, and later it was granted to him—imprisonment in his villa, under guard by a representative of the Inquisition, which gave the Inquisition the means to keep an eye of him. He wouldn't dare mention Copernicus and his theory again, because if he were found guilty of "relapsed heresy," he could be put to death. He was even denied permission to seek the treatment that might have prevented his complete blindness.

It is possible, notes Stillman Drake, that this was the intention of the Inquisitors all along. Even if they had wanted to, they could not acquit Galileo. The Holy Office itself had brought the charges against the aged scientist, and finding Galileo innocent would be the same as admitting that he had been falsely charged by the Holy Office, an accusation as serious as that of heresy. Meanwhile, as Galileo suffered at the hands of Rome, his *Dialogue* was a hit on the black market.

Galileo was forced to live the remainder of his long life in shame and increasing blindness. But even as his sight was leaving him, he *still* had the energy to gaze intelligently at the sky that had led to his great fame and subsequent downfall. Was it teasing him with yet another new find? Although he was in great pain, he observed that the Moon did not always present the same face toward the Earth. It seemed to oscillate ever so slightly, sometimes to the right or left, sometimes up and down. This effect is now called libration.[20]

One wonders how, after the utter indignity suffered at the hands of his former friend Pope Urban VIII, coupled by the death of his daughter the next year, he had any drive left in him to think, to write, and to publish. It is something of a miracle not only that he did, but that his last work, written while imprisoned in his home, was his best and most important: *Mathematical Discourses Concerning Two New Sciences, Relating to Mechanics and to Local Motion*. Once again Galileo chose the style of a dialogue, but this time the subject was uniform acceleration, in which a body in motion receives equal increases in velocity per equal units of time. He also gave for the first time the correct theory of the motion of projectiles. He was a pariah of the Church, but the steady stream of visitors in Galileo's final years might have given him the impression, subsequently shown to be correct, that history would treat Galileo kindly.

There is a famous story that as Galileo walked out of the great hall during which his sentence was pronounced, he muttered *Eppur si muove*, words that defy the Inquisition, words that state that the Earth does move. If he did, it was to the credit of the people standing nearest to him that they pretended not to hear. But the year before his death, knowing full well that the anger of Pope Urban VIII had not even begun to fade, he still wrote to Fortunio Liceti, a man who could have gotten him into more difficulty, "Now your honor may see what a hard task it will be for those who want to make the Earth the center of the planetary circles. A place which could be, as it were, a center to all planets except the Moon befits more the Sun than anything else."[21]

GALILEO IN 1992

Some decades after Galileo's death in 1642, the Church permitted a complete republication of Galileo's *Dialogue*. This was not an admission that it had erred, but it was a tacit wink of the eye in favor of the Copernican theory. In 1978 shortly after Pope John Paul II took office in 1978, he announced that the Church should reevaluate the wrong it had done to Galileo. He felt that a new look at the Galileo trial, which poisoned relations between science and faith, was vital. At last, in 1992, three and a half centuries after it sent Galileo to his villa prison, the Church declared that what Galileo had done was not a crime. "The error of the theologians of the time" noted Pope John Paul II, "when they maintained the centrality of the Earth, was to think that our understanding of the physical world's structure was, in some way, imposed by the literal sense of Sacred Scripture."[22]

Strictly speaking, it was not Galileo's discoveries that sparked his troubles, it was the conclusions that Galileo drew from his discoveries. However, interpretation is an integral part of the discovery process.

This chapter has dealt with possibly the most serious consequence ever to have been faced by a discoverer, and I wish to bring some perspective to the story with the following anecdote. One evening in 1993, my cat and I witnessed a space shuttle reentering the atmosphere in a bright train of orange fire that lit up the southern sky. It was so bright that the cat watched the orange meteor move from west to east across the sky. Can it be said that the cat discovered the glow in some way? Without any means for the cat to interpret what he saw, I doubt it. At the risk of seeming trite, I contend that if a discoverer does nothing with a find, then it is not a discovery. The cat did nothing. Galileo first recorded his observations, then published them in detail, then used them to show that Earth was not the center of the universe, and finally he faced the consequences of his interpretations. In the conduct of his own life and beliefs, Galileo carried the process of discovery full circle.

SIX

1656:
CHRISTIAAN HUYGENS
It Is Surrounded by a Ring

Annulo cingitur, tennui, plano, nusquam coherente, ad eclipticam incli-nato. (It is surrounded by a thin, flat, ring, nowhere touching, inclined to the ecliptic.)

—a deciphered anagram by Christiaan Huygens, 1656

Four years before Galileo went on trial, Christiaan Huygens was born in Holland, a different land both physically and intellectually from seven-teenth-century Italy. Both Galileo's and Christiaan's fathers had a healthy interest in science and supported their sons' interest in nature. But unlike Galileo, Christiaan Huygens was never made to suffer for his scientific quest.

Christiaan's father, Constantijn Huygens, a man of eclectic interests, was sec-retary to Prince Frederick Henry of Orange. He enjoyed poetry so much that he penned verse in seven languages. He especially liked the poems of English poet John Donne, with whom he was close friends. It is probably no accident that Donne tackled the new philosophy of Copernicus so elegantly in his *First Anniversary*, which appeared only one year after Galileo's discovery of the moons of Jupiter:

And new philosophy calls all in doubt,
The Element of fire is quite put out;
The Sun is lost, and th' Earth, and no mans wit
Can well direct him where to looke for it.[1]

Young Christiaan Huygens enjoyed a childhood home with frequent guests from many nations, inspired dinner conversations, and opportunities to expand his mind in many directions. It was a childhood that today's youth would envy. "The world is my country," Christiaan would say, and "science my religion."[2] One of those special visitors was the French philosopher René Descartes, the father of modern philosophy. Descartes would steer the young Huygens's thinking in an interesting cosmological direction. In his *Principles of Philosophy* of 1644, Descartes proposed a universe filled with particles in constant motion and collision, resulting in geometric shapes and vortices. He also established the idea that a particle at rest in the void of space will stay at rest, and one in motion will stay in motion. Descartes filled the universe, however, with particles swimming through a void where their motions were constantly being changed by collisions with other particles. And although Huygens later modified Descartes's vision of particle motion, particularly when studying light, he always accepted Descartes's basic idea that the universe was filled.

Another visitor was Antony van Leeuwenhoek, who did not actually invent the microscope—Galileo made one in 1610, years before Leeuwenhoek was born—but who did make 270 microscopes and use them in the study of medicine. Huygens was fascinated with the idea that lenses specially placed could enormously magnify minute things. He thus built microscopes and used them to study the movements of human sperm cells, thereby helping to understand for the first time the process of human reproduction. By exploring microbes in water, he developed the idea that microscopic life forms could float through the air and alight in water, even water that had been boiled. In that sense, Huygens, with Leeuwenhoek, had ideas that later led to the theory that germs spread disease two centuries before the work of Louis Pasteur.

In May 1645, the sixteen-year-old Huygens entered the University of Leiden to study the seemingly unrelated fields of law and mathematics. He followed a course in law, most likely to please his father, who hoped that he would follow a diplomatic career, and in mathematics because he loved it. Even here his fortunate list of contacts did not desert him. Huygens's mathematics professor, Frans van Schooten, carefully helped him develop into a careful mathematical thinker.

In 1647 Huygens transferred to the University of Breda to continue law, but the optics of the microscope and telescope kept pulling him in a different direction. By 1651, at the young age of twenty-two, Huygens published his first book, *Theoremata de quadratura hyperboles, ellypsis et circuli*, a work that allowed him to put the geometrical skills he had learned from van Schooten to good use. This book was Huygens's clear indication that he wanted to follow a career in science.

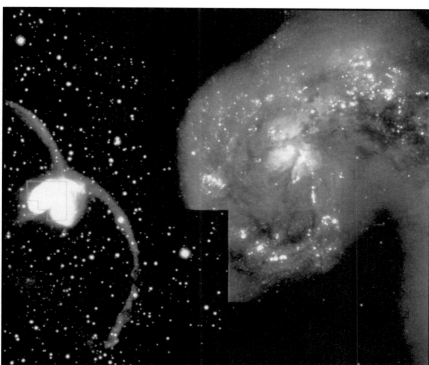

Colliding Galaxies NGC 4038 and 4039 are exquisite in this Hubble Space Telescope photograph, but they can be spotted through a small telescope. (*NASA/HST image*)

David and Wendee Wallach-Levy at the Jarnac Observatory entrance. (*Photo by Joan-ellen Rosenthal*)

A Herbig-Haro Object illustrates a star in its process of formation; Bart Bok studied these exciting stars. *(Hubble Space Telescope image/ NASA)*

We do not know exactly when Comet Shoemaker-Levy 9 was born, but we do know that its last major fragment disappeared from the solar system early on July 21, 1994, in an explosion recorded by the *Galileo* spacecraft. *(NASA/JPL image)*

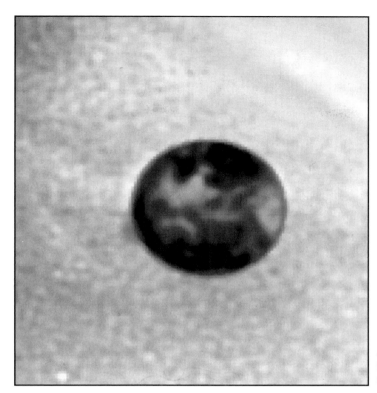

A Volcano on Io.
(Hubble Space Telescope image/NASA)

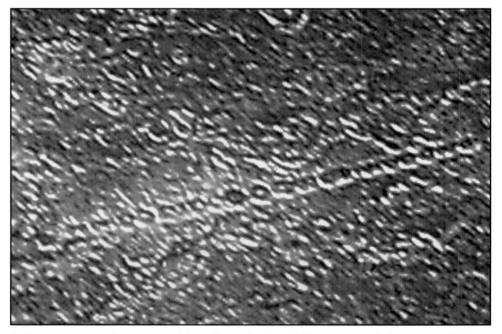

A chain, or catena, of craters caused by a Shoemaker-Levy 9-like impact on Jupiter's moon Callisto. *(Voyager Photograph. Courtesy NASA/JPL)*

Gravitational lensing, as seen through the Hubble Space Telescope. The bright cluster of galaxies is lensing the light from more distant galaxies, which appear as streaks around the cluster. *(NASA/HST)*

Comet Levy, 1990. *(Photo by David H. Levy through the eighteen-inch telescope at Palomar)*

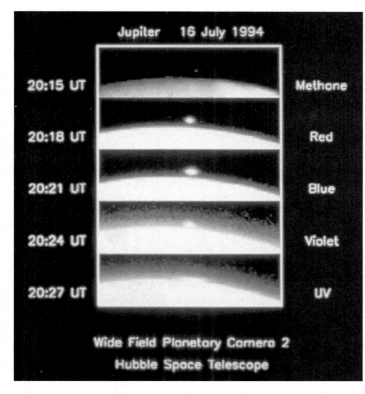

The impact of Comet Shoemaker-Levy 9's Fragment G on Jupiter. This Hubble Space Telescope image shows the sequence of events when the first fragment crashed on July 16, 1994. *(NASA/HST)*

NGC 4414, a spiral galaxy photographed through the Hubble Space Telescope. *(NASA/HST)*

A Shoemaker-Levy 9 press conference. *(NASA photo)*

Clyde Tombaugh photographed next to author David Levy's first telescope. *(Photo by David H. Levy)*

Tycho Brahe, in typical court attire, ready to begin a night's observing. *(Courtesy Lick Observatory)*

Galileo Galilei, possibly around the time of his trial in 1632. *(Courtesy Yerkes Observatory)*

William Herschel, discoverer of the solar system's seventh planet. *(Courtesy Lick Observatory)*

John Couch Adams, co-discoverer of Neptune, many years after his find. *(Courtesy Lick Observatory)*

Urbain Jean-Joseph LeVerrier, co-discoverer of Neptune, is honored with a statue in front of the Paris Observatory. *(Courtesy Yerkes Observatory)*

Stonehenge, December 1976. *(Photo by David H. Levy)*

Henrietta Leavitt at her desk at Harvard College Observatory. *(Courtesy Yerkes Observatory)*

HUYGENS'S GREAT YEARS: 1655–1656

Huygens's young life changed abruptly as a result of five months in Paris in 1655. He was astonished and delighted with the quality of intellectual life there. He spent time with lens grinders with a view to improving the quality of the telescopes he could use. As a result, he developed an improved method of grinding and polishing lenses. To improve the process, Huygens, working at times with his brother, Constantijn, used large tools made of copper and brass to grind the lenses for his telescopes. These tools were usually more than twice the diameter of the lenses he was grinding.[3]

The first result was a fabulous new telescope with a focal length of twelve feet with which he discovered, in March 1655, Saturn's largest moon. Now named Titan, this moon, almost as big as Mars, is the second largest in the solar system after Ganymede. It has a dense atmosphere. In 2004, a probe, named *Huygens* after Titan's discoverer, will sail into Titan's atmosphere, exploring what lies beneath its dense clouds.

Huygens also used this telescope to try to resolve a major enigma concerning Saturn that had lasted almost half a century. The problem began when Galileo, in July 1610, turned his telescope to Saturn, which had recently become visible after several months of conjunction with the Sun. "I have discovered a most extraordinary marvel," he told the Grand Duke. "The planet Saturn is not one alone, but is composed of three, which almost touch one another and never move nor change with respect to one another."[4] Through his telescope he saw two ear-like appendages that, over several years, grew smaller and vanished. Then after a season in which the appendages were completely absent, they returned—small at first, then larger.

Huygens was only twenty-six when he peered through his new telescope at Saturn. He did not see any appendages, but instead, the view through his telescope hinted strongly that Saturn was surrounded by a ring. Reluctant to announce the discovery at first, he translated his message into a simple Latin sentence, and published it as an anagram like this:

AAAAAA CCCCC D EEEEE G H IIIIII LLLL MM NNNNNNNNNN OOOO PP Q RR S TTTTT UUUUU. When worked out, the anagram read *"Annulo cingitur, tennui, plano, nusquam coherente, ad eclipticam inclinato,"* which, translated into English, means "It is surrounded by a thin, flat, ring, nowhere touching, inclined to the ecliptic."[5]

In 1656 Huygens devised a new type of eyepiece, now known as the Huygenian eyepiece. Through his new telescope that same year, Huygens discovered the four stars in the center of the Orion Nebula. The next years were very active in a variety of areas: Huygens invented a spirally balanced spring that was used in watches for centuries, and he even contributed to the law of probability by studying how dice work in a game. He invented a "magic lantern," the grandfather of modern slide projectors. He invented also what he called the "gunpowder engine," a precursor of the steam engine.

Most of these incredible feats took place before he celebrated his thirtieth birthday in 1659! Using a telescope with a lens two and one-third inches across and a focal length of twenty-three feet, in that year Huygens confirmed the ring and published his discovery in *Systema Saturnium*. Eventually, the Huygens brothers built a 210-foot telescope. Christiaan was the first to recognize how the role of the steadiness of the atmosphere or "seeing," plays in how well a telescope performs. He often warned observers not to blame their telescopes too quickly, when poor seeing may be the cause of fuzzy images.[6]

Unlike other telescope makers of that era, Huygens also did not use long wooden tubes for his telescopes. Instead, with his "aerial telescope" he mounted the objective lens, or object glass, into a small iron tube that was in turn mounted on a platform near the top of a pole. The platform could be raised or lowered like a flag on a flagpole. The lens formed the image near the ground, where the observer could maneuver the eyepiece. The eyepiece and object glass were connected with a thread. Huygens must have become quite expert at somehow managing to bring the glass and eyepiece into alignment.

CASSINI AND TIDAL FORCES IN THE SOLAR SYSTEM

The discovery of Saturn's ring was astonishing, and it quickly explained Saturn's changing shape over its twenty-nine-year orbit of the Sun. Huygens explained that every fourteen years, when Earth passes through the plane of Saturn's rings, the rings appear edge-on and almost disappear.

No life of Huygens is complete without mention of Giovanni Domenico Cassini, a contemporary who was born in northern Italy in 1625. Giovanni Cassini correctly interpreted the rings as being made of "a swarm of tiny satellites."[7] As a young man he also made extensive observations of Jupiter, Mars, and Venus, increasing his reputation by discovering the rotation periods of Jupiter and Mars, as well as accurate tables of the motions of Jupiter's moons. In 1669 Cassini moved to Paris where, known as Jean Dominique Cassini, he became superintendent of the construction of the Paris Observatory, a magnificent building that was mostly completed by 1671.

The Paris observatory was unique in that it was planned as a public observatory. In 1671 Cassini discovered Saturn's moon Iapetus, and the next year he found a second moon, Rhea. His signature discovery came in 1675, when he detected a dark marking within Saturn's ring. This turned out to be a division between two of Saturn's rings, and is now named Cassini's Division in his honor. With a better telescope in 1684 he discovered two more moons, Dione and Tethys. In 1693, his revised tables of the orbits of Jupiter's moons set a new standard for accuracy.

Saturn's rings are one of the most dramatic examples of tidal pull in the solar system. This swarm isn't really satellites, but tiny particles of ice, the remains of long-gone moons or comets that got too close to the planet and broke apart because of Saturn's tidal pull. Centuries later, the planet Jupiter offered two other great examples of tidal pull. One is a volcano on Jupiter's moon Io that was discovered when *Voyager 1* sailed by in March 1979. As Io swings around Jupiter, its orbit is affected by its sister moons, Europa and Ganymede. Jupiter's tidal pull stretches Io during each orbit. As the interior changes shape every day, it stays warm, allowing volcanoes all over that moon spew out sulfur. The other Jovian example of tidal pull happened in July 1993, when Comet Shoemaker-Levy 9 passed within 15,000 miles of Jupiter's cloud tops and disrupted catastrophically into twenty-one pieces.

There is a famous example of tides right here on Earth at the eastern end of Canada's Bay of Fundy. There at mid-tide, a flow of water equal to the combined currents of all the rivers on Earth thunders through the Minas Channel into the Minas Basin. For six hours the water flows into the basin like a bathtub overflowing. As the first quarter Moon hung above us in the sky when I visited there, a *million million* gallons of water poured into the Minas Basin. I found it incredible that the Moon, from a distance of about 240,000 miles, was responsible for the tremendous noise in the channel below. After a few hours the flow slowed, then stopped. For a half hour or so the water around Cape Split was still. Then the huge basin started to empty, flowing *in reverse*, like a movie running backward.

THE PENDULUM CLOCK

Huygens's discovery of Saturn's ring system made him famous, but during that same creative period of his life, he would make an even more important contribution. In 1656 Huygens was also becoming very interested in clocks, in large measure because his astronomical observations required a more precise measurement of time than was available then. His thinking thus turned to applying the motion of Galileo's pendulum as a regulator of a clock. The findings of his clock research, which resulted in the first pendulum clock, appeared in his *Horologium* (1658) and in his 1673 book *Horologium Oscillatorium.* The latter book offered the derivation of the formula for the time of oscillation of a simple pendulum. Far more than a discussion of pendulum clocks, the book contained Huygens's theory on the mathematics of curvatures and the laws of centrifugal force in a circular motion.

The next part of Huygens's life was spent partly in Paris, partly in the Hague. By 1662, in the Hague, he continued working on projects as varied as telescope lenses, clocks, and even on vacuum air pumps. The next year, during a visit to London, Huygens was given his first major honor, a fellowship in the Royal Society. Back in Paris the same year, Louis XIV offered him a very good pension to continue his scientific work. It was the king's goal to set up an *Académie Royale des Sciences*. After some delay Huygens returned to Paris in May 1666 and became a founding member of this French Academy of Sciences. He remained in Paris for much of the rest of his life. While there, he spent most of his time researching clocks, but he also developed his wave theory of light.

PROBLEMS IN PARIS

By 1670, Huygens's virtually charmed life hit some real problems with a breakdown in his health. After he recovered he worked on his *Horologium Oscillatorium*, a book as well known for its opening pages as it was for its science. At its publication in 1673 Huygens dedicated it to France's Louis XIV but worded it rather strangely, almost as if it was in memory of the still living king. Since France was at war with Holland at the time, Huygens was sharply criticized for the awkward wording. Not letting this bother him, Huygens continued to live in Paris where, in 1678, while experimenting with the mineral calcite, he discovered the polarization of light by noticing that light is doubly refracted when it passes through calcite.

In 1681 he became very ill again and returned to Holland. While he was there he learned of the death of his patron in France, Jean-Baptiste Colbert, adviser to Louis XIV. Four years later Louis revoked the Edict of Nantes which had allowed Protestants like Huygens to live peacefully in France. These two changes made Huygens feel unwelcome at the thought of returning to France, and he decided to stay in Holland.

In 1689 Huygens made one final visit to England, where he met with Isaac Newton, whose *Principia Mathematica* had just been completed and published. Although the two scientists highly respected each other, Huygens did not agree with Newton's theory of gravity because it did not have a mechanical explanation. In 1690 Huygens published, in French, his *Discours de la cause de la pesanteur*, a discourse on the cause of gravity, which explained gravity as a mechanism involving the vortices of his old mentor Descartes. Huygens also developed the theory that light travels in waves, a theory still partially accepted today. Huygens's principle states that every point on a light wave front is itself a source of new waves of light. His *Traité de la Lumière (Treatise on Light)* was published in 1690. The book included elegant descriptions of the nature of reflection and refraction of light; it was based entirely on his principle of secondary wave fronts. Huygens wrote a third book at that time, the fanciful *The Celestial Worlds Discover'd: Conjectures Concerning the Inhabitants, Plants and Productions of the Worlds in the Planets.* In it, he dreamed up people whose shapes are vastly different from ours. It was a lively book, masking his own feelings of sadness as his health declined.

Huygens died in the Hague in 1695. In his memory a southern constellation, *Horologium*, is named in his honor. The clock is a small constellation, east of Archernar. Since its brightest star is fourth magnitude, this constellation is hard to identify. It is one of the star patterns set aside by Nicolas Lacaille, who intended this constellation to honor the invention of the pendulum clock by Christiaan Huygens. Lacaille's original name for this constellation was *Horologium oscillatorium*, the title of Huygens's most famous book. It is a thoroughly justified honor.

SEVEN

1682:
EDMOND HALLEY

Discovery by Calculation

> *Wherefore if according to what we have already said it should return again about the year 1758, candid posterity will not refuse to acknowledge that this was first discovered by an Englishman.*
> —Edmond Halley, *Astronomical Tables*

Halley's comet brings generations together. Every seventy-six years, grandparents stand outdoors with their grandchildren, peering into the night as Halley's Comet moves majestically, peering back at us on its once-in-a-lifetime inspection tour of Earth. As comet hunter Leslie Peltier noted, this comet has been around since prehistoric times, presided over the defeat of Attila the Hun in 451 C.E., and terrified the warriors of the Norman conquest of England in 1066. In 1456 Halley's comet frightened Pope Calixtus III into ordering days of prayer. "In 1607," Peltier wrote, "it was watched by both Shakespeare and Kepler and I like to think that it was also seen by Captain John Smith and Poca-hontas in the frontier skies of Jamestown. On its following trip around in 1682 the comet was observed by Halley himself, who probed into its periodic past and bequeathed to it an honored name that it can bear with pride throughout the solar system."[1] In 1835 the comet presided over the birth of Mark Twain, and it was there when he died in 1910.

For the comet's most recent visit in 1986, the International Halley Watch coordinated many thousands of observations, and the Soviet Union, Europe, and Japan sent a flotilla of spacecraft to meet it. As part of the Halley Watch program I started a venture called "Project 2061" in which teachers and children were given directions to find the comet, and hopefully they were inspired to remember their diaphanous traveler when it returns in 2061. When the comet returns again in 2134, we may have to duck, figuratively at least, for viewers then will enjoy the comet's closest pass to the Earth in 1,300 years. Peltier asks what wonders of technology we will have to study the comet in the future, or "will man himself prove periodic? Will the Huns be back again?"

EDMOND HALLEY OF LONDON

As we have seen, Tycho Brahe established that comets did not form within our own atmosphere, as Aristotle believed, but that they travel out in space farther than the Moon. That was in 1577. More than forty years later, Galileo, Horatio Grassi, and Mario Guiducci were still arguing the point, and there the matter stood, until 1682, when a major comet dominated the sky over England and attracted the attention of a young Englishman named Edmond Halley.

Born November 8, 1656, Halley was the son of a soapmaker and salter, and like other subjects of our book so far, Halley's father, though not a scientist, strongly encouraged Edmond to develop an interest in science and nature. Halley followed Huygens's path by publishing at a young age; he was only twenty when his first paper appeared while he was at Oxford, but unlike Huygens, Halley did not feel he had to spend his time in a college classroom. Halley's life actually turned sharply in 1676. After the publication of his paper he dropped out of Oxford and headed south to the island of St. Helena, the island that, more than a century later, would serve as Napoleon's place of exile after the Battle of Waterloo.

Halley sailed to that south Atlantic island because he wanted to chart the southern sky and compile a catalog of stars in the southern hemisphere. He observed frenetically through his telescope over the next thirteen months, sailed back to England, and published his *Catalogue of the Southern Stars.* His professors at Oxford were so delighted with Halley's accomplishment that they allowed him to re-enter Oxford, and later awarded him a master's degree without requiring him to take the exams.

Halley did not share the stiff-upper-lip personality of his fellow scientists; he was warm and open. "One can imagine that Newton was at his best," writes science historian Donald Yeomans, "when left alone in his study to focus his prodigious intellect on mathematical constructs. The stern John Flamsteed, Astronomer Royal, was at his best in the Greenwich Observatory making precise and careful position measurements of celestial bodies. No doubt Halley was at his best discussing new scientific ideas with friends at a favorite coffee house."[2]

THE COMET TRIO OF 1680, 1682, AND 1683

For humanity's understanding of the nature of comets, the first part of the 1680s was the most crucial time in history. It began on November 14, 1680, when Gottfried Kirsch, while looking at the sky through his telescope, became the first human to discover a comet using a telescope. From that day until March 19, 1681, when Isaac Newton observed it through a different telescope, the comet was observed almost constantly; it played a major role in modernizing humanity's understanding of comets. As a first step, astronomer Georg Dörffel observed the comet, measured its positions in relation to background stars, and used these positions to describe the comet's orbit as a parabola, with the Sun at its focal point. Next came Astronomer Royal John Flamsteed, who observed the comet in November 1680 as it approached the Sun, then predicted that it would reappear in December, moving away from the Sun. He was naturally delighted when the comet reappeared on schedule and raced from the Sun in his prescribed direction.

While all this excitement was going on, Halley was also observing this magnificent comet, quietly noting its changing positions. But the quiet that followed the comet's departure was not to last long. In 1682 a bigger and brighter comet appeared, marching across the sky, its tail dwarfing the earlier one. When yet a third comet came by in 1683, Halley's special interest was piqued. He wanted to understand their orbits, and he suspected that he could picture them if the calculations were based on the highly regarded observations made by Flamsteed. Some time passed before, in 1687, he wrote to Isaac Newton suggesting that the great scientist use his own mathematics to calculate these orbits. By 1695, Halley had learned how to make the difficult calculations himself, at least for the last of the three comets. The result, he wrote Newton, was that Halley's and Flamsteed's observations of the comet could best be fitted to a parabolic orbit, like the originally calculated orbit of the comet of 1680. Only three weeks later Halley wrote to Newton again, commenting that he had just recalculated the orbit for the 1680 comet using only Flamsteed's observations, and found that the positions best fit into a long ellipse rather than a parabola.

It was the Comet of 1682, however, that really stirred Halley's curiosity. His calculations, based on comet positions that he and others derived, indicated that its orbit was identical to the orbits of the comets of 1531 and 1607. Could the three comets really be a single comet that was traveling to the outskirts of the solar system and then returning to the vicinity of the Sun? The only problem with that idea, Halley easily noticed, was that although the intervals were close, they were not identical. The time between 1531 and 1607 was seventy-six years; but only seventy-five years separate 1607 from 1682. Although the comet of 1683 was exactly seventy-six years from 1607, its orbit was not even close to those of the others. Halley begged Newton to obtain Flamsteed's careful observations of this middle of the three comets, especially those taken in September 1682. "I am more and more confirmed," he wrote, "that we have seen that Comet now three times, since ye Yeare 1531." [3] Now seriously interested in the possibility that one comet was returning again and again, he determined the paths of twenty-one other comets that appeared between 1337 and 1698.

In 1696, Halley gave a now-famous presentation to the Royal Society of London. He was certain that the comets of 1607 and 1682 were one and the same. A month later he addressed the Royal Society a second time, demonstrating that one of the comets of 1618— the cluster that had ushered Galileo back into public scientific debate—was in a parabolic orbit that brought it closer to the Sun than Mercury.[4]

The Comet of 1618 presentation was Halley's last for a while, for the increasingly well-known astronomer had other interests he wanted to follow. First he left London to become deputy controller of the English Mint, then in 1698 he became a sea captain on the *Paramore Pink*. The three voyages he took were probably the first strictly scientific journeys in history. During his first two trips across the Atlantic, he made careful observations of the differences between magnetic north and true north in an attempt to see if these variations could help with the old and serious problem of determining longitude at sea. (The Earth has a magnetic field with north and south poles, but these poles change over time, and are not at the same place as the planet's geographical poles. The Earth's true poles form the ends of the axis around which the planet rotates.) This problem of longitude couldn't be solved in this way; there was no consistent result to help with longitude. It was solved, in fact, with John Harrison's clocks, which could retain accurate time even while aboard a swaying ship, by the latter half of the eighteenth century.[5] In 1701, Halley sailed again, this time to produce the first accurate chart of the English Channel tides.

Returning from these sailings, Halley finally completed his *Synopsis of the Astronomy of Comets*, publishing it at last in 1705. Its highlight is a single table giving the orbital elements for twenty-four comets that appeared between 1337 and 1698. He showed that all these comets were bound to the Sun either in parabolic orbits, or, as he really believed, very long elliptical orbits that would eventually bring them back to the vicinity of the Sun. The most important part of the 1705 document was its addition of the comet of 1531 to the pair of comets in 1607 and 1682. Halley was now convinced that all three were the same comet. He also suspected that three earlier comets, 1305, 1380, and 1456 were also apparitions of this one comet; it turned out that he was wrong with the first two but absolutely correct to include the comet of 1456.[6]

What had, at long last, allowed him to add these earlier comets was his solution of the discrepancy of seventy-five versus seventy-six years by factoring in the effects of Jupiter's gravity. Halley had discovered that Jupiter, and the other large planets, had a considerable effect on a comet's path. Over the centuries, the effect of "planetary perturbations" would be measured to have varying effects on other comets. For some, the effect would be negligible. For Comet Shoemaker-Levy 9, the effect would be catastrophic.

Halley had had his work cut out for him four centuries ago. "He really had the benefit only of the cometary orbits he computed," notes Brian Marsden, "not so much the records of the comets themselves at roughly 76-year intervals." It was far more difficult to link Halley's Comet with comets in the more distant past. Although a bright comet appeared in 1301, he thought that the comet he was studying should have returned four years later.[7] It was not until the nineteenth century that celestial mechanicians like J. Russell Hind were able to connect earlier apparitions to Halley's comet, and early this century Cowell and Crommelin used ancient records to confirm the comet's visitations as far back as 240 B.C.E.

In a revised edition of his *Astronomical Tables,* Halley capped his work with a prediction that his comet would return at the end of 1758 or early in 1759. "Wherefore if according to what we have already said it should return again about the year 1758," Halley concluded triumphantly, "candid posterity will not refuse to acknowledge that this was first discovered by an Englishman."[8]

HALLEY'S DEATH AND LEGACY

Halley had to leave his prediction to posterity, for he knew it was a good bet that he wouldn't live to age 103 in order to see his comet. He died in 1742, after a full life of service to astronomy that included a long stint as England's Astronomer Royal. Even his revised prediction was published after his death. With each refined calculation, Halley seemed to grow less confident on the return date of the comet. Donald Yeomans notes that in Halley's first prediction in 1705, he wrote in Latin that "I shall venture confidently to predict its return in 1758." When he translated that into English later that year, it appeared as a far more modest "I dare venture to foretell." Ten years later Halley hedged even more with "I think, I may venture to foretell"; and in his last writings he meekly said "if . . . it should return again about the year 1758."[9] One of today's most experienced orbit computers, Yeomans believed that as Halley grew older he became more certain of just how complicated the calculation of comet orbits really is when the effects of the planets are taken into account. After all, he could predict all he wanted, but his comet was out there beyond the orbit of Neptune, and only that ball of ice and dirt could know precisely where it was and when it would return.

Although his fame rests mostly on the comet that bears his name, Halley's career boasted other major achievements. From his many observations of stars separated by decades during his long life, he discovered that stars do change their positions relative to each other. He also was the inspiration for several expeditions around the world to see Venus in two of its rare transits as the planet passed in front of the Sun in 1761 and 1769. This was all in response to his suggestion that observations of these transits could allow a precise calculation of the distance between Earth and the Sun. We will remember Halley again during the forthcoming transits of Venus that take place in 2004 and 2012.

The story of Halley's life does not end with his death, for his comet was still inching its way back for another visit. At the start of 1758, many astronomers frantically searched for it on paper by calculating where in the sky it might be, while others scanned the sky with telescopes. As the army of calculators raced to provide a place to look for the comet's return, a farmer and amateur astronomer near Dresden, Johann Georg Palitzsch, went outdoors on a frigid Christmas night in 1758. Palitzsch thought he spotted something with his unaided eye low in the southwestern sky, a dim fuzzy spot in the constellation of Pisces. With some agitation he tried to set up his telescope, but the cold and his nerves cost him precious minutes. Finally his telescope was ready and he set it on the fuzzy object in time to draw its position relative to some nearby stars in his field of view. Halley's comet had returned, precisely when Halley said it would!

This achievement inspired the following generation to continue the mathematical search for other possibly returning comets. In 1765, Nicolas-Louis de Lacaille first referred to it as Halley's Comet.[10] Five years later, Anders Johan Lexell computed that a comet discovered by Charles Messier in 1770 had a period of only five years. But there was more: This strange comet, Lexell's calculations showed, had come close to Jupiter just before its approach to the Earth, and afterward it approached Jupiter a second time. In this later encounter, Jupiter batted the comet right out of the "ball park" of our solar system; never again would Lexell's comet return to the vicinity of Earth or Sun.

In this book's Preface, we explored the work of Johann Franz Encke, who linked the apparitions of comets that appeared in 1786, 1795, 1805, and 1819 to fit the orbit of a single comet that would next return in 1822. Encke was right, and the comet which now bears his name has by far the shortest orbital period of any comet, a brief three and one-third years. Halley's comet thus stood alone for some sixty years as the only comet proven to be periodic, a tribute to the rare genius of the man whose name is forever attached to it.

EIGHT

1760:
CHARLES MESSIER

The Comet Hunter of Cluny

> *What caused me to undertake the catalogue was the nebula I discovered*
> *above the southern horn of Taurus on 12 September 1758, while observing*
> *the comet of that year. . . . This nebula had such a resemblance to a comet,*
> *in its form and brightness, that I endeavored to find others, so that*
> *astronomers would not confuse these same nebulae with comets just begin-*
> *ning to shine.*
>
> —Charles Messier, *Connaissance des Temps*, 1801[1]

I t is never easy to point to a particular event or happening that led someone to
a career in discovery. For Charles Messier, however, it is probable that two
comets, the comet of 1744 and Halley's Comet fifteen years later, played crucial
roles in his early decision to search for comets. On December 9, 1744, Dirk
Klinkenberg from Holland discovered a faint naked-eye comet. Four days later
Philippe Loys de Chesaux discovered it from Switzerland. The comet brightened
until it was more brilliant than the crescent Moon. For three days in early March
that comet sported a spectacular system of as many as eleven tails.

This comet couldn't have failed to attract the attention of a fourteen-year-old
Charles Messier. Born in 1730 in Badonviller, a small village in Lorraine in the
south of France, Messier was the tenth of twelve children. Compared to the
people whose lives we have explored in earlier chapters, Messier spent his childhood in poverty; his father passed away when Charles was only eleven years old.
On July 25, 1748, an annular eclipse of the Sun had a profound influence on
Messier's curiosity about astronomy. Three years later, having completed his
schooling and seeing no chance of improving his life in Lorraine, Messier set out

76

for Paris to find his fortune. He arrived in Paris with nothing, and presented himself to a Parisian astronomer named Delambres, who noted later that the young man had "hardly any other recommendation than a neat and legible handwriting and some little ability at draughtsmanship."[2] He soon met Nicholas Delisle, who had recently set up a marine observatory in an old fifteenth-century building then known as the Hôtel de Cluny. Delisle hired Messier to keep the records of observations at the observatory. As clerk at the observatory, Messier did everything he could to follow his goal of observing the night sky.

MESSIER AND HALLEY'S COMET

The year 1758 was a crucial one for the twenty-eight-year-old Messier. Brimming with enthusiasm about the probable and imminent return of Halley's Comet, Messier decided that he would be the first to spot it. That year many people were searching for the comet. Delisle, who wanted his young assistant to succeed in spotting it, prepared for Messier two charts using positions he had calculated. Messier began searching in the middle of 1757 and searched right through 1758. He also observed his first comet that year. On January 21, 1759, Messier finally glimpsed Halley's Comet a month after Johann Palitzsch. However, even in early 1759 Messier had not received this news:

> The whole day was very fine and without cloud [he wrote]; in the evening I went over the sky with the telescope, keeping the limits of the two ovals drawn [by Delisle] on the celestial chart which was my guide. At about six o'clock I discovered a faint glow resembling that of the comet I had observed the previous year: it was the Comet itself, appearing 52 days before perihelion!
>
> There is cause to presume that if M. Delisle had not made the limits of the two ovals so restricted, I would have discovered the comet much earlier, while it had a greater elongation from the Sun.[3]

Thrilled with his discovery, which he called one of the most important finds in the history of astronomy since it showed that comets do return as they orbit the Sun, Messier quickly told Delisle. For some reason never understood, Delisle ordered Messier not to announce his achievement in any way. It wasn't until Messier recovered the comet a second time, after it already had rounded the Sun, that Delisle finally permitted him to announce the discovery. That announcement took place on April 1, 1759; coming as it did months after Messier's preperihelion observations, other French astronomers ridiculed Messier. The triple disappointment of not being allowed to announce his find, then having it questioned, and finally being beaten at his own game by Palitzsch, possibly encouraged him to begin his own search for comets.

THE FIRST COMET HUNTING PROGRAM

Messier didn't have long to wait to assuage his disappointment. Beginning what is now known as the first successful search for new comets, Messier discovered his first comet in 1760. He must have felt he was on his way to repairing his reputation. However once again, Delisle did not permit his assistant to announce it right away. As aggravating as this must have been, Messier never complained, at least not publicly. Shortly after this comet discovery, however, Delisle decided to retire. The Cluny Observatory was sort of left to Messier, in a somewhat backhanded way. Although he was allowed to use the telescope, he was not allowed to use Delisle's apartment, now empty, nor was he given any increase in salary. His position as clerk remained.

One year after his first comet find, Messier observed the transit of Venus that Halley realized was so important. He discovered his second comet on September 28, 1763, his third on January 3, 1764 (with his unaided eye), and a fourth on March 8, 1766. By this time, few remembered the early difficulties of Messier's comet hunting career. Messier was made a member of the Royal Society of London in 1764. He considered it only a matter of time before his own country would honor him with membership in its prestigious *Academie Royale des Sciences*. Messier really wanted this honor even before the English one was given to him. "I solicited this associate membership," he wrote about his 1763 attempt; "I presented to the Academy my journals of observations since 1752."[4] But membership was denied him; the fact that Messier was *just* an observer, and not a scientist, kept coming back to haunt the comet hunter despite his growing successes.

In 1769 Messier discovered his fifth comet. By now quite experienced both in how to report his comets and how to make the most of them, he wrote of his discovery to many people, including King Frederick II of Prussia. On September 2, 1769, Frederick the Great replied

> The letter which you wrote to me on the 15th August last to inform me of the discovery you have made of a new comet in the constellation of Aries, greatly pleased me and I am very grateful to you. As I have passed on this information to the department of mathematics at the Berlin Academy of Sciences, Sieur de la Grange, I would be delighted if you would correspond with him more fully on this subject; and in this, I pray that God will keep you in His holy care.
>
> Frederick.[5]

Messier achieved what he wanted out of that correspondence; membership in the Berlin Academy of Sciences. But it was his sixth comet, about which we have already read, that assured his fame and respect. Though Anders Lexell, over the centuries, has received more credit for calculating its strange orbit than

Messier has for discovering it in 1770, the comet's fate became the subject of much scientific discussion. It had a five-and-a-half-year period, then a close approach to Jupiter expelled it from the inner part of the solar system. At last, the scientists at the *Academie Royale des Sciences* were prepared to recognize Messier's accomplishments, and that year took the rare step of electing a mere observer of the sky to their ranks as an Academician.

MESSIER'S CATALOGUE

As the first serious and systematic comet hunter, Messier learned about the sand traps of the sport in time to warn the rest of us. Fuzzy objects, galaxies, nebulae, and clusters that look like comets, which comet hunter Leslie Peltier called "comet masqueraders," are all over the sky.[6] At the end of 1758, when Messier found a fuzzy patch near the star Beta Tauri that he at first thought was a comet but which never moved as a comet would, he began to build a catalogue of these deceivers.

Although Messier was primarily a comet hunter, he is infinitely more famous today for his catalogue. The first entry in his catalogue, now called Messier 1 or M1, is the Crab Nebula because it looks somewhat like one. Although he never realized it, Messier 1 was probably the most interesting thing ever to cast light into his telescope. It is the remnant of the supernova, a near-total destruction of a star, that was observed on July 4, 1054, to be as bright as Venus. A petroglyph in northern New Mexico depicting a star near a crescent Moon may be a record of this supernova, but although the date seems right, this relation has yet to be confirmed. In only 700 years that explosion had produced a huge expanding shell of gas bright enough that Messier could easily see it. In 1969, observers using a thirty-six-inch-diameter telescope at Steward Observatory in Arizona found an object at its center, a spinning neutron star called a pulsar, that spins thirty times per second.

As the numbers of objects grew, Messier published three versions of his catalogue. Of the forty-five objects of his first catalogue that appeared in 1774, no more than eighteen were his own discoveries. Messier didn't discover all the objects in his catalogue, certainly not M45, the Pleiades. But he did list, in his catalogue and in his journals, 110 objects. Observing all of them has become the goal of many amateur astronomers. In 1962 I began my own Messier hunt with a single observation of the Pleiades. In the spring of 1967, using a larger telescope than I had in 1962, I finished the list while observing from my grandfather's cottage at Jarnac Pond, Quebec. In March 1983 I observed them all again, all but M30, during the course of a single night.

GOLDEN YEARS

After twelve years of struggle, Messier had finally achieved his goal. Even King Louis XV praised his accomplishments, calling him "the ferret of comets." Messier kept on searching. On April 1, 1771, he discovered a seventh comet with his naked eye, and, at long last, was appointed Astronomer of the Navy, with a great increase in salary. He then discovered his eighth comet on October 13, 1773. But now Messier was getting some competition. Jacques Montaigne, a druggist in Limoges, France, snared his first in 1772, the comet that in 1826 became known as Biela's Periodic Comet or P/Biela. This comet apparently split in two in 1846, returned as two comets in 1852, and only as a meteor storm in 1872. By 1781 Pierre Méchain had also joined the contest. But Messier was still searching hard, and he found his ninth comet on October 27, 1780.

The next year, 1781, was almost totally taken up with three events. One took place in England on March 13, and its story, the discovery of a new planet, occupied Messier's energy for several months. (The full story is told in the next chapter.) On April 13, 1781, a month after William Herschel's discovery of Uranus, Messier found his hundredth cometlike object. Although he intended to round off his list with what is now known to be a galaxy in Virgo, at the last minute he added three more objects suggested by his friend Méchain, which Messier did not have the time to observe himself before the publication deadline of his catalogue.

The third major event happened on November 6, 1781. Messier had become close friends with President de Saron of the Paris parliament, a politician also versed in the science of celestial mechanics. He had asked for observations of all Messier's comets so that he could calculate their orbits and predict their future courses across the sky. The two men enjoyed each other's company, both scientifically and socially. On November 6 de Saron and his family asked if Messier would like to join them in a visit to a beautiful oriental garden at the Parc Monceau, with many small pagodas, castles, and interesting passageways.

Thoroughly enjoying the intricate walks, Messier got curious about a small open door at the end of one of them. It was dark on the other side. Thinking this was just part of the excitement of the park, he stepped through and fell some twenty-five feet, landing on ice. Someone had left the door open to the park's storeroom of ice. Messier was critically injured, and was lucky to have survived thanks to someone having noticed the opened door. With great difficulty Messier was rescued, but he had broken his arm, a leg, several ribs, and split his scalp.

Messier was bedridden for almost a year, and when his leg was not properly set the surgeon had to refracture it. He gradually was able to resume his activities in time to observe a transit of Mercury, as it passed in front of the Sun, on November 12, 1782.

THE COMET HOAX

On May 14, 1784, Messier started searching the constellation of Vulpecula for a comet that had been reported to him that day by a man known as the Chevalier Jean Auguste d'Angos. Messier was puzzled why, after a thorough search over several nights while still recovering from his injuries, he could not find this comet, since it had clear positions and even an orbit that the chevalier had computed from fourteen observed positions.

D'Angos was a physician by profession and a chemist and astronomer by avocation. Although the comet was not seen by anyone else at the time, its mystery kept its record alive for years, long after Messier's own role in the story had ended. In 1820, Johann Encke, still working on the orbit of his own comet, decided to investigate the chevalier's comet in hopes of finding it on some subsequent return. Encke made an astonishing discovery of his own: The orbit that the chevalier had calculated did indeed fit the positions he had observed, but only when Encke multiplied the distance from the Earth to the comet by ten. The chevalier had invented an orbit, then wrongly plugged the imaginary comet into it, all which wasted the time of Messier and others. "D'Angos had the audacity," Encke wrote, "to forge observations he had never made, of a comet that he had never seen, based upon an orbit he had gratuitously invented, all to give himself the glory of having discovered a comet."[7]

The chevalier worked his magic twice more, in 1793, and in 1798. While we will never know if the 1793 comet was real, no one else saw it. The 1798 comet he "observed" transiting the Sun in broad daylight. It turned out that Joseph Lalande's textbook *Astronomie* suggested that the Sun happened to be at the exact position of the descending node (a specific point in the orbit) of the Comet of 1672 on January 18, 1798. However, that figure turned out to be in error by sixty degrees, a sixth of the way around the sky. Hevelius, the Polish astronomer who had found the comet back in 1672, would have been amazed to see his comet as part of a hoax.

One final note about the chevalier, this hapless physician who, we hope, was better as a doctor than he was at either of his hobbies. He was not much better at chemistry than he was in astronomy. One night, while doing experiments with phosphorus, he set his observatory at Malta on fire and it burned to the ground.[8]

MESSIER'S FINAL YEARS

For Messier, the chevalier incident occupied only a few days of his life as he was completing his recovery from his accident. By 1785 he announced his full reentry into observing with his discovery of a real comet, his tenth, on January

7, 1785. Méchain was observing the same night and co-discovered the same comet. On November 26, 1788, Messier discovered his eleventh, this one in Ursa Major. By this time political events in France were taking over the lives of everyone in Paris. The following summer the Bastille was stormed and set afire, and the revolution that followed saddened and frightened Messier. In 1791 the Paris Parliament, of which Messier's friend de Saron had been president at one time, was disbanded, and de Saron was imprisoned.

By early 1793, the revolution had grown savage thanks to Robespierre and his Reign of Terror. On January 21, 1793, Louis XVI was guillotined. Nine months later, on September 27, 1793, Messier found his twelfth comet in Ophiuchus, one he was able to observe until it got too close to the Sun in October. Like so many times before, he informed de Saron, this time by smuggling a note into prison. De Saron attempted to calculate an orbit using the positions Messier supplied.

On December 29, 1793, Messier searched the morning sky and found his comet close to the position de Saron had predicted for it. Messier wrote of de Saron's last success and hid his note in a newspaper which he was able to smuggle to the prisoner. On April 20, 1794, just three months before the end of the Reign of Terror, de Saron was guillotined.

Although Messier survived the revolution and the reign of terror that followed it, his pension, which de Saron had arranged for him after his accident, was gone, and the acclaimed astronomer was as penniless as he had been as a child in Lorraine. Messier did have help from astronomer Joseph Lalande, who managed to care for him and others who were being sought after by the revolutionaries. The dark recesses of the Paris Observatory served well as a hiding place for several scientists, priests, and others who feared for their lives.

The revolution finally ended, and with the calming down of tensions, France steadied its course. The Academy of Sciences was reopened, with Messier as a member, and possibly more important, he was made a member of the Board of Longitude. The board was responsible for examining all serious attempts to find a solution to the problem of determining longitude, either through the sky or using mechanical clocks.[9] Messier still searched the skies at his observatory at Cluny, where in 1798 he found his thirteenth comet not far from the Pleiades.

Messier's wife passed away that same year. There is a well-known but unverified story from his Russian friend La Harpe that Messier found out about a discovery by Montaigne while Messier was mourning his wife's death. A friend embraced the grief-stricken man to say, "I am so sorry." Messier glared at his visitor. "Alas," he said, "Montaigne has robbed me of my comet!" Quickly realizing his faux-pas, Messier tried to recover. "Poor woman," he muttered. No doubt his friend agreed.[10] According to Messier scholar Kenneth Glyn Jones, the story is apocryphal since Montaigne's comet was back in 1772, not 1798. However, another comet discoverer, Eugene Bouvard, did find a comet that December. Whether true or an exaggeration, the story probably reflects Messier's intensity and passion for comets.[11]

Messier found his fourteenth and last comet in 1801, the same comet that was Jean-Louis Pons's first discovery. Pons would go on to find at least thirty comets, a record until Carolyn Shoemaker found thirty-two comets almost two centuries later.

Despite this long and successful life, Messier lived modestly at Cluny. Although he rarely, if ever, spent time at the Paris Observatory, he did occasionally use its nineteen-foot-long refractor. He was mostly satisfied with his own telescopes, of which a reflector with a mirror just over seven inches diameter was the largest. In 1806 he received the Cross of the Legion of Honor from Napoleon. In return, he dedicated the comet he had discovered in 1769 to Napoleon, who had been born the same year. This is the same comet he wrote Frederick the Great about, but now it was part of a piece of writing entitled *Sur la Grande Comete qui a paru a la Naissance de Napoleon le Grand.* It is interesting that this somewhat astrological description came from the pen of the great Messier.[12]

On April 11, 1817, Messier died after a long life at the age of eighty-seven. It was not until 1921 that some later movement occurred with his catalogue. That year Camille Flammarion, a French popularizer of astronomy, found some notes about another object in Messier's personal copy of the catalogue. That object was the Sombrero Galaxy in Virgo, and was added to the list as M104. In 1947 Helen Sawyer Hogg, a well known Canadian astronomer, suggested that Messier was well aware of three additional objects found by Méchain. Not long afterward Owen Gingerich at Harvard suggested that two galaxies in Ursa Major that were noted in the original copy should be M108 and 109. Kenneth Glyn Jones suggested the addition of the fainter companion to M31, the Andromeda Galaxy, since Messier had observed it in 1773 and drawn it in 1807. Thus, Messier's catalogue lists 110 clusters, nebulae, and galaxies. Not all those objects belong there; M102 seems to be a misplaced position for 101, and M40 is simply two stars. Regardless, this catalogue remains the number one list of "deep sky objects" in the world. Amateur astronomers constantly refer to it, search for its members, and try to see them all.[13]

In 1986, while visiting the Paris Observatory, I perused one of Messier's observing logs. As I studied his notes, his descriptions of the telescope he used, his complaints about the weather, and his comments about what he was looking at, two hundred years between us seemed to vanish. Not that much has changed; reflecting the same concerns and joys, his notes could well have been my own. In that notebook, protected under glass, lies the spirit and soul of one of the truly great observers. May it inspire many generations of people to love the sky and observe it well.

NINE

1781:
CAROLINE AND
WILLIAM HERSCHEL

*On Tuesday the 13th March, between ten and eleven in the evening, while
I was examining the small stars in the neighborhood of H Geminorum, I
perceived one that appeared visibly larger than the rest; being struck with
its uncommon magnitude I compared it to H Geminorum and the small star
in the quartile between Auriga and Gemini, and finding it so much larger
than either of them, suspected it to be a comet.*

—William Herschel, 1781[1]

Imagine a clear, crisp English night. In his small back yard, a forty-three-year-old man is peering through the eyepiece of a six-inch-diameter reflecting telescope. This scene is all too familiar to hundreds of thousands of amateur astronomers who enjoy the evening hours with their telescopes. Except for the fact that the six-inch telescope, at ten feet, is longer than most modern telescopes, this could have happened last night. Only it didn't. The year was 1781, and the amateur astronomer was Friedrich Wilhelm Herschel, who with his younger sister, Caroline, was about to become the most famous brother–sister astronomical team in the history of astronomy.

Born in Hanover, Germany, in 1738, Frederich Wilhelm Herschel was the son of a successful musician, and as a youngster he learned to play the horn, harp, and organ. When he moved to England in 1757, he changed his first name to William. In England Herschel's highest ambition was to establish his reputation in music. To get from place to place to give recitals, he bought himself a horse and he frequently rode across the fields in all kinds of weather. His frenetic schedule did not let up after a hard day: "During all this time," he wrote, "though it afforded not much leisure for study, I had not forgot my former plan, but had given all my leisure hours to the study of languages. After I had improved myself sufficiently in English, I soon acquired the Italian, which I looked upon as necessary for my business. I proceeded next to Latin, and having also made considerable progress in that language, I made an attempt of the Greek . . . but soon dropped the pursuit of that as leading me too far from my other studies, by taking up too much of my leisure."

These "other studies," Herschel goes on, were in music. "The theory of music being connected with mathematics, had induced me very early to read in Germany all that had been written upon the subject of harmony; and when, not long after my arrival in England, the valuable book of Dr. Smith's harmonics came into my hands, I perceived my ignorance and had recourse for other authors for information, by which I was drawn from one branch of mathematics to another."[2]

Two years after he moved to England, the twenty-two-year-old Herschel took a vacation and a concert tour through Italy. When he reached Genoa near the end of his tour, he found he had run out of funds for his return trip to England. Desperate to get home, he put on a bizarre recital— holding a harp and two horns, he played all three. The concert was a great success and he raised enough money to return home.

A BOOK OF ASTRONOMY

In August 1772, Herschel urged his younger sister, Caroline, then twenty-two, to join him in England. Caroline was twelve years younger than her brother, and had known him for only the first seven years of her life. She wanted to join him, and the two Herschels rapidly became close. The following May, while walking through a bookshop, William found a book that looked interesting. The resulting purchase led to a major change in the direction of his life. "May 10," he wrote in his diary: "Bought a book of astronomy and one of astronomical tables."[3] The books captivated him, and soon his impressive energy was redirected to the heavens. With some lenses he bought later, he built a four-foot long telescope. "With this," he wrote, "I began to look at the planets and the stars. It magnified 40 times. In the next place I attempted a 12 feet one and contrived a stand for it. After this I made a 15 feet and also a 30 feet refractor and observed with them."

Herschel learned two things with these refracting telescopes: one, they gave impressive images of the Moon, planets, and objects beyond; and two, they were long and difficult to use. Moving a thirty-foot-long tube across the sky was not easy, especially if a breeze kept the tube from staying straight enough to allow the light to travel from the object glass to the eyepiece. Herschel next thought of using reflector telescopes of the design invented by Isaac Newton in 1668 by putting a concave mirror at the bottom of a tube and an eyepiece near the top.

In early September 1773, Herschel completed a two-foot-long Gregorian-type reflector, in which a concave secondary mirror sends light through a hole in the main mirror to the eyepiece. "This was so much more convenient than my long glasses," he wrote, "that I soon resolved to try whether I could not make myself such another. . . . I was, however, informed that there lived in Bath a person who amused himself with repolishing and making reflecting mirrors."

When Herschel arrived, however, he found that the optician was preparing to retire from his work. After what must have been an interesting conversation, the older man offered Herschel "all his tools and some half-finished mirrors, as he did not intend to do any more work of that kind. . . . About the 21st October I had some mirrors cast for a two-feet reflector." Following the custom of the day, Herschel did not use glass but a form of speculum metal which consisted, he wrote, "of 21 copper, 13 tin, and one of Regulus of Antimony, and I found it very good, sound white metal."[4]

So thoroughly did Herschel immerse himself in his zeal for telescopes that he and Caroline needed a larger house in Bath. Caroline caught the astronomy bug at the same time, as well, joining him enthusiastically in his work and not minding all the telescopes in various stages of construction strewn about. The backyard of their house was set up as an observing site. About the size of a small modern backyard, it had room for several small telescopes.

MARCH 13, 1781

Only eight years after he fell in love with the stars, William was as good at using telescopes as he was at making them. He was also in the midst of a systematic survey of the sky with a seven-foot-long reflector which had a 6.2-inch-diameter mirror. On March 13, 1781, while looking through his telescope, he saw something fairly bright and large. It could not be a star, since all the stars he knew about were points of light. To perceive this new object as a disk of light rather than a point required a good telescope, very good eyesight, and an ability to discern the details at the limit of vision. With eight years of experience behind him, Herschel had perfected the art of seeing faint details through his telescope. This art served him well that night. "Seeing is in some respects," Herschel wrote later, "an art which must be learnt. To make a person see with such a power is nearly the same as if I had been asked to make him play one of Handel's fugues upon the organ. Many a night have I been practicing to see, and it would be strange if one did not acquire a certain dexterity by such constant practice."[5]

Thinking it must be a comet, he investigated at higher magnification, and then he announced his discovery. News of this peculiar "comet" spread swiftly to Charles Messier, who observed it from Cluny at every opportunity. Messier was stumped by both its ultra-slow movement from night to night and its shape. He wrote to Herschel in the late spring of 1781,

> I am constantly astonished at this comet, which has none of the distinctive char-acters of comets, as it does not resemble any one of those I have observed, whose number is eighteen. . . .
>
> I have since learnt by a letter from London that it is to you, Sir, that we owe this discovery. It does you the more honor, as nothing could be more difficult than to recognize it, and I cannot conceive how you were able to return several times to this star—or comet—as it was absolutely necessary to observe it sev-eral days in succession to perceive that it had motion. . . .
>
> For the rest this discovery does you much honour; allow me to compliment you for it. I should be very curious, Sir, to learn the details of this discovery, and you will oblige me if you will be so good as to inform me of them.[6]

By the end of May, the constellation of Gemini, with its new and slow-moving object, was hidden in the glare of the Sun. But Anders Lexell, the same celestial mathematician who had calculated the orbit of Messier's 1770 comet, continued to work on the observed positions to try to determine its orbit. On August 31, 1781, he excitedly announced that it had an almost completely circular orbit. Furthermore, it was orbiting the Sun at a very great distance—far beyond the orbit of Saturn, it never gets closer to the Sun than sixteen times the Earth's distance to the Sun. Now it was known: Herschel's object did not look much like a comet because it was not a comet. It was a planet beyond Saturn, the first to be discovered in historic times.

The discovery was a sensation, but Herschel insisted that it was no accident. He was proud of his discovery but prouder still of the thoroughness of his survey that led him to the new world. Since no new major planet had been found in recorded history, Herschel's discovery was unheard of, and it quickly earned him the Royal Society's highest honor, the Copley Medal: "Your attention to the improvement of telescopes has already amply repaid the labour which you have bestowed upon them;" the citation read, "but the treasures of the heavens are well known to be inexhaustible. Who can say but your new star, which exceeds Saturn in its distance from the sun, may exceed him as much in magnificence of atten-dance? Who can say what new rings, new satellites, or what other nameless and numberless phenomena remain behind, waiting to reward future industry?"[7] Almost two centuries later, in 1977, the planet Herschel discovered, Uranus, passed in front of a star. The star blinked out several times before Uranus passed in front of it, then it repeated the performance the same number of times after. Thus a set of rings was found encircling Uranus. More rings were found when

Voyager 2 sailed past Uranus in 1986 as part of its odyssey through the outer solar system. The planet is known to have twenty-one satellites, two of which Herschel discovered himself.

All did not go well during that historical year of 1781, however. On August 11, while preparing to cast a mirror for a proposed telescope thirty feet long, the casting mold started to leak and the mirror cracked. Herschel remelted the pieces of metal and tried again, but this time the liquid metal started to drop into the fire. The molten metal poured onto the stone floor. As his assistant raced out of the building, Herschel, angry, tired, and hot, fell, out of frustration, onto a pile of bricks.[8]

In the summer of 1782, King George III invited Herschel for an audience. An avid amateur astronomer, George had established his Kew Observatory near the present site of the Kew Botanical Gardens in London. He had ordered the construction of the observatory in 1768 so that he could observe the transit of Venus the next year. The grounds, building, and telescopes were most impressive, and would soon include a ten-foot-long reflector made by Herschel. The meeting took place at Windsor Castle, on July 2, 1782. Herschel was delighted with their observing session that evening. "My Instrument gave a general satisfaction," he wrote to his sister Caroline; "the King has very good eyes and enjoys Observations with the Telescopes exceedingly."[9] That particular observing session went well for William Herschel. The king offered him a lifetime pension of £200, and thus, from that point on, Herschel gave up performing music for a living and devoted his time exclusively to astronomy.

In 1786 the Herschels moved to a place between Bath and London named Slough, which happened to be within view of Windsor Castle, the distant walls of which was a good target on which to test the quality of his mirrors. Anxious to improve his telescopes beyond the seven-foot reflector, Herschel constructed a twenty-foot-long telescope whose top hung from a mast with ropes. He later built a "large 20-foot" telescope, equipped with a mirror 18.7 inches in diameter. The mounting was more efficient; it consisted of a large wooden framework that could not only lift the telescope but also swing it in azimuth. The telescope and mounting were not covered by any building, but were left exposed to all kinds of weather. Because the mirrors were made of metal, not glass, they were subject to tarnishing, and the only way to cure that problem was to repolish the surface of the mirror. Thus Herschel had a collection of four mirrors for the twenty-foot telescope, so that when one was being repolished, another would see the stars.

THE DISCOVERY OF INFRARED RADIATION

Herchel's observing was active during daylight hours as well. He devised an optical wedge that, when inserted in place of the telescope's secondary mirror, reduced the amount of sunlight that reached the eyepiece by 96 percent. In 1800 he placed a prism into the light path of his telescope and divided the Sun's light into its spectral colors. Once he had produced the spectrum, he used a thermometer to record the temperature change at different wavelengths. Possibly by accident, but more likely because of his customary thoroughness, Herschel did not stop measuring at the red end of the spectrum, and with great surprise he noticed that the thermometer recorded a rise in temperature, even though there was no visible sunlight there. Puzzled as to why this should happen, Herschel concluded that the Sun's heat and light were fundamentally different properties.

Not stopping with this aspect of the Sun, Herschel suspected that our home star might vary in light like some of the distant stars he had been observing. Wondering if these variations could result in changes in the Earth's climate, he searched for a way to measure subtle changes in climate. He thought he found one in the price of wheat. By comparing the price from good years to bad, changing with climate, Herschel thought he might be able to measure the variation of heat output from the Sun. Although this idea now seems naive, it was a brilliant attempt to study the interaction between Earth and Sun.

Herschel made other observations of the Sun, particularly with sunspots. "In the year 1783," he wrote, "I observed a fine large spot, and followed it up to the edge of the sun's limb. Here I took notice that the spot was plainly depressed below the surface of the sun; and that it had very broad shelving sides."[10] His observations led to an imaginative theory: Only the atmosphere of the Sun is very hot, but the interior is cool and dark. Sunspots are gateways, or windows, to that comfortable interior. Herschel went on, "The sun, viewed in this light, appears to be nothing else than a very eminent, large, and lucid planet." Not only that, but "it is most probably also inhabited, like the rest of the planets, by beings whose organs are adapted to the peculiar circumstances of that vast globe. . . ."[11]

Herschel was obviously wrong in this interpretation, as he was wrong in his idea that the Sun's heat and light were fundamentally different. His cold Sun theory went nowhere. But without knowing it, his simple thermometer experiment with the solar spectrum turned out to be a seminal discovery of modern science. William Herschel had clinched the discovery of infrared radiation. In the two centuries since his finding, astronomers have studied the sky, from the ground and through space satellites, in the infrared portion of the spectrum.

THE COMETS OF CAROLINE

William Herschel's self-discipline and enthusiasm rubbed off on Caroline, who eagerly set about assisting her brother's astronomical work. Around 1780 Caroline began hunting for comets using a six-inch-diameter reflector her brother had built for her. By 1783 she had discovered the beautiful spiral galaxy in Sculptor now known as NGC 253, but her real goal was to follow Messier and discover comets.

Caroline's first success came on the night of August 1, 1786. Her brother was away in Germany at the time, and it is possible that she was using the time of his absence to do her own searching. She found the comet as bright as the Dumbbell Nebula, which is No. 27 in Messier's catalogue. While she was certain that the object was a comet, a hazy sky prevented her from confirming its motion. Thus she passed a tense and exciting twenty-four hours hoping the sky would be clear enough the following evening. The sky did cooperate, and she noticed that the comet had moved.

It goes without saying that her brother was thrilled, though probably not surprised, with the news. But one of the official observatory comments, though polite, was somewhat condescending: "Let us hope the best and that it is approaching the earth to please and instruct us and not to destroy us, for true Astronomers have no fears of that kind. . . . I would not affirm that there may not be some astronomers so enthusiastic that they would not dislike to be whisked away from this low terrestrial spot into the higher regions of the heavens by the tail of a comet."[12]

Four days before Christmas in 1788, Caroline discovered her second comet in Lyra. William was with her this time, and he described it as "a considerably bright nebula, of an irregular form, very gradually brighter in the middle, and about five or six arcminutes in diameter."[13] Although the comet's orbit was considered to be parabolic, meaning that the comet would not return, a century and a half later, in 1939, Roger Rigollet, of Lagny, France, found a comet that turned out to be Herschel's comet on its first return since discovery. Thus, this comet is now called Herschel-Rigollet, and it has a period of about 150 years.

As Caroline's comet finds became more routine, she was awarded a small income. She had a banner year in 1790. She found her third comet on January 7, a difficult find which only three days later came too close to the Sun to be seen. On April 18, 1790, she discovered her fourth comet, this time in the predawn sky in the constellation of Andromeda. Comet Herschel No. 5 turned up on December 15, 1791. A few days before she discovered it, the comet had made a relatively close pass by the Earth. We have already seen how, on November 7, 1795, Herschel's sixth comet was later determined to be the first observed return of Encke's Comet. By the time she found her last comet in 1797, England's most famous woman astronomer was known throughout Europe for her "eccentric vocation" that included a log of her discoveries lovingly entitled *Bills and Receipts of My Comets*. Her seventh and final find, Bouvard-Herschel-Lee, was a bright, naked-eye object at discovery and fairly close to Earth. The Royal Astronomical Society bestowed on her its Gold Medal in 1828.

THE HERSCHEL TEAM

Despite her personal success with comet hunting, Caroline Herschel always believed that her greatest contribution was in assisting her brother. They remained close friends and colleagues even after William, in May 1788, married Mary Baldwin Pitt, a widow with a considerable fortune. "His wife seems good-natured," a visiting novelist wrote about her, and added that "astronomers are as able as other men to discern that gold can glitter as well as stars."[14] The Herschels had a son.

With two large grants totaling £4,000 from King George, William built a mammoth forty-foot-long telescope with a forty-eight-inch-diameter mirror. The huge wooden scaffolding next to Herschel's house in Slough was clearly visible from the road near their home. Its triangular arrangement of wooden poles, ladders, and planks stretched fifty feet above the ground. On the first night of its use in 1789, Herschel discovered Mimas and Enceladus, two of Saturn's closest moons. Although this was his biggest telescope, it was very difficult to use, and its large mirror tarnished quickly. The eyepiece could be reached only by standing on a high platform and shouting orders to the assistant, usually Caroline. The telescope attracted many visitors, especially King George, who once said to the Archbishop of Canterbury, "Come, my Lord Bishop, I will show you the way to Heaven!"[15]

Carolyn always feared for her brother while he was high atop one of his telescopes. "My brother began a series of sweeps when the instrument was yet in a very unfinished state. . . . The ladders had not even their braces at the bottom; and one night, in a very high wind, he had hardly touched the ground before the whole apparatus came down."[16] But the person who ended up injured was Caroline herself, not William. She had a bad accident on the telescope one night when her ankle was struck by a large hook hidden in snow. As the story goes, one night William had asked her to move the telescope in a particular direction. The telescope didn't move, and when he heard nothing from his sister, he repeated, "Lina, move the telescope!" But a large hook used for pulling the telescope had snagged her ankle. "I'm hooked!" Carolyn yelled back. As soon as she was almost recovered from this injury, she returned to assist her brother in his work. Caroline was not alone with accidents. Giuseppe Piazzi, who discovered the first asteroid, Ceres, on January 1, 1801, fell over a device called a rack bar while visiting the telescope.

As William grew older, he stopped making regular searches for new nebulae with the twenty-foot in 1802. He became very ill in 1807, and though he recovered, his health remained delicate. He stopped using the twenty-foot altogether after the summer of 1814. Around this time William's son, John, after studying mathematics, decided to follow the path of his father and aunt. In 1820, the year of King George III's death following years of intermittent mental illness, John Herschel restored the twenty-foot telescope and started making regular observations with it. William must have been delighted that the telescope was continuing to see productive use. When he died two years later, Caroline left England for Hanover. Although she continued to receive honors, she lived alone there, and obviously lonely, for twenty-six years. "You will see what a solitary and useless life I have led these 17 years," she wrote, "all owing to not finding Hanover, nor anyone in it, like what I left, when the best of brothers took me with him to England in August, 1772."[17]

HERSCHEL FAMILY LEGACY

Even though he was alive at the same time as Messier, William Herschel had a great deal more interest in building telescopes than Messier did. Possibly because of lack of funds, Messier did not improve much on the small instruments available to him at Cluny. By the end of his life William had made more than a thousand telescopes. Some of these telescopes survive; one is on display in Chicago's Adler planetarium. He was proud of his telescopes, but prouder still of what he had learned to see with them.

Because Herschel learned how to build much larger telescopes and use them, he took Messier's list and extended it to some 2,500 deep-sky objects. He wrote,

> It would be hard to be condemned, because I have tried to improve telescopes and practised continually to see with them. These instruments have played me so many tricks that I have at last found them out in many of their humours and have made them confess to me what they would have concealed, if I had not with such perseverance and patience courted them. I have tortured them with powers, flattered them with attendance to find out the critical moments when they would act, tried them with specula of short or long focus, a large aperture or a narrow one; it would be hard if they had not been kind to me at last.[18]

Over their many years of observing, William and Caroline measured the rotation rate of Saturn—ten hours—by repeatedly measuring the time it took markings on the planet to cross its face. Since the markings on Saturn are less distinct than those of Jupiter, this was no small achievement. Moving out beyond the planets, they measured the positions of each component, relative to the other, of more than 800 double stars. In 1803 William Herschel concluded that some of these doubles actually revolve around each other. This achievement required a lifetime of work, since the orbital periods of these stars often take decades.

In 1834, John Herschel moved the twenty-foot reflector to the Cape of Good Hope in South Africa. From the sale of many of his father's telescopes and from his father's inheritance, he had become wealthy and could spend all his time in observing and in study. His years in South Africa with his family and his telescope were comfortable ones. During these years he discovered and measured many previously unseen clouds of gas and clusters of stars in the southern sky.[19] When he returned from his survey in 1837, John found the forty-foot telescope in a dangerous condition. With some reluctance he decided to dismantle the telescope and scaffolding. Years later during a thunderstorm, lightning hit a tree which completed the destruction of the telescope. All that now remains of William Herschel's great forty-foot telescope is the bottom stub from the tube and two of its mirrors.

When the *Voyager* spacecraft sailed by Saturn and its moon, Mimas, in 1981, it found a crater so large that the impact that created it must have almost split the moon apart. The large crater is named Herschel, in honor of the team that contributed so much to our understanding of the universe.

TEN

1846:
ADAMS, LEVERRIER,
AND THE SCANDAL
OVER NEPTUNE

You see, Uranus is long way out of his course. I mean to find out why. I think I know.

—John Couch Adams,
Cambridge University student, circa 1841[1]

Is this an hour

For private sorrow's barren song,
When more and more the people throng
The chairs and thrones of civil power?

A time to sicken and to swoon
When Science reaches forth her arms
To feel from world to world, and charms
Her secret from the latest moon?

—Alfred, Lord Tennyson, *In Memoriam*, 1850,
four years after Neptune and its moon Triton were discovered[2]

If all the discoverers in this book could be gathered in one room to share their stories, I think that they would find their experiences divided into two categories. Some, like Huygens and the Herschels, would tell exciting tales of finding new rings and new worlds. Others would shake their heads as they recounted the controversies that swirled around them as a direct result of the work they did and the discoveries they made. John Adams and Urbain Leverrier, the two celestial mechanicians who pointed the way to Neptune, are squarely in the corner of controversy.

Their story might have started with Galileo's first discoveries regarding Jupiter in 1610. Besides plotting the positions of the moons, Galileo noted where distant stars were in the field of his telescope. Neptune happened to be in conjunction with Jupiter at the time, and it is likely that Galileo actually sketched it without knowing what it was. Almost two centuries later Joseph Lalande, the French astronomer who had provided valuable protection for Messier during the French Revolution, was checking fields of stars that he had mapped two nights earlier through his telescope at the Paris Observatory. On the night of May 10, 1795, he found a mistake. An eighth magnitude star was not in the same place it had been. Surprised at his apparent carelessness, he corrected his chart and moved on. Lalande never revisited that field, and never noticed that his wrongly plotted star would continue to move slowly, elegantly, across the background of stars.

How history might have changed if Lalande had gone back to that field one more time. His misplotted star was Neptune, the eighth major planet in our solar system, and it would not be found for another fifty-one years. The story of the discovery of Neptune would then have been an *observational* story; instead it became more of a *mathematical* one that began when astronomers found an old sketch of Uranus that Tobias Mayer, an observer from Göttingen, made on the night of September 26, 1756—a full quarter century before Herschel's discovery. Astronomers eventually collected twenty-three "prediscovery" observations of Uranus—that is, drawings that were made by astronomers before Herschel found the planet. These are all drawings by people who, like Lalande with Neptune, didn't know onto what they had stumbled. These old observations allowed astronomers to see that Uranus was not following the orbit that it should be. Finally, a prediscovery observation by John Flamsteed turned up in 1690. This critical piece of data allowed astronomers to see where Uranus was more than a full orbit before its official discovery. Flamsteed's observation confirmed that, clearly, something was wrong with the new planet's orbit.

In 1834 the amateur astronomer Rev. Thomas J. Hussey wrote to George Airy, who two years later would become the Astronomer Royal of England. Hussey had a proposal: Could some distant body, another planet perhaps, be affecting the orbit of Uranus? Although the problem was a concern to astronomers, Hussey's proposal helped to define it better.

Hussey wasn't alone in his idea. Two celestial mechanicians, Anders Johann Lexell and Giuseppe Boscovitch, experienced in plotting the orbits of planets and comets, decided that Uranus's orbit could probably be explained by one and possibly more distant planets. Airy was aware of the problem; two years earlier, he had reported that the planet was almost half a minute of arc off its course. There had been other ideas; one was that Uranus was having trouble pushing its way through the Cartesian fluid of outer space far from the Sun. (Many astronomers still subscribed to René Descartes's view that space was filled with a cosmic fluid, rather than being the vacuum we now know it to be.) Another interesting idea was that Uranus had recently been struck by a comet, which significantly changed its orbit. That theory, it turns out, returned two centuries later to explain why Uranus has such an extreme tilt that it rolls through the sky on its side. In any event, Airy's response to Hussey was the first of a series of bad decisions on the part of the Astronomer Royal. Conceding that the gravitational influence of another body might indeed be at work, Airy thought that such a body would be too hard to find.

JOHN COUCH ADAMS

When Hussey made his suggestion of a new planet, John Couch Adams was a teenager. Seven years later, he was a twenty-three-year-old Cambridge University student who was thinking about the Uranus problem after reading Airy's report on it. On July 3, 1841, Adams decided to spend some time after graduation investigating the Uranus problem, a plan "of investigating, as soon as possible after taking my degree, the irregularities in the motion of Uranus which are as yet unaccounted for; in order to find whether they may be attributed to the action of an undiscovered planet beyond it; and if possible thence to determine the elements of its orbit, etc. approximately, which would probably lead its discovery."[3] The orbit could be solved, he thought, by the presence of another world, and after his graduation he used his vacation time to calculate where such a world might be found in the sky.

Adams made his first attempt at an orbit calculation in 1843. It was a rudimentary orbit—completely circular, at twice Uranus's average distance from the Sun. Considering the distances of other planets from the Sun, Adams thought it reasonable to suspect that the new world would be at that distance. Adams talked at length to John Challis, his former professor of astronomy at Cambridge. Challis obliged by obtaining additional positions of Uranus from Airy's file, and delivering them to Adams.

In September 1845, Adams attempted a solution of a new orbit, based on more observations of Uranus. Instead of a circular orbit, he proposed a slightly elliptical one and sent the solution along to Challis, who forwarded Adams's work to Airy. Once again Airy, now Astronomer Royal and director of the Greenwich Observatory, voiced interest but incredibly did not encourage a search with a telescope. Adams worried, with much justification it turned out, that time was running out, and so he pushed matters along by trying to visit Airy at the Greenwich Observatory in late October. During this "visit" he knocked on Airy's front door twice; Airy was out the first time, and at dinner the second, so he left a summary of his work and his calling card with the butler. Adams left frustrated and angry.

Airy was interested in Adams's work, and he replied by letter his understanding of the situation. "I am very much obliged by the paper of results which you left here a few days since, shewing the perturbations on the place of *Uranus* produced by a planet with certain assumed elements. The latter numbers are extremely satisfactory: I am not enough acquainted with Flamsteed's observations about 1690 to say whether they bear such an error, but I think it extremely probable."[4]

It was what Airy wrote in his concluding remarks that upset Adams. "I should be very glad to know whether this assumed perturbation will explain the error of the radius vector of Uranus. This error is now very considerable."[5] Adams thought that the radius vector of the orbit was not an important factor, and he was impatient at what he perceived to be Airy's lack of understanding. Thus, he didn't answer Airy's letter for more than a year.[6] Had Adams not been so angry, and a little more persistent, Neptune might have been found earlier.

Why no one else mounted a search in England on the basis of Adams's prediction is a difficult question to answer. England boasted several telescopes large enough to search thoroughly in that area; in fact most amateur scopes six inches in diameter or larger could have detected it. The simplest answer is that other than Airy, Challis, and Adams, no one knew to look.

Airy was a man of great genius, but he was difficult. Under his regime, observing assistants underwent an extremely difficult three-day routine involving, on the first day, twenty-one nonstop hours on a measuring instrument called the transit circle. Day two was, in comparison to the other days, a holiday, consisting of a small amount of computing, but day three followed with all-day computing and all-night observing. Then day one of the next cycle would begin at six the next morning![7] Is it possible that Airy's preoccupation with the observatory's routine work prevented him from paying more attention to something out of the ordinary, like Adams's prediction, and then deploying his observatory's telescopes to the search for Adams's planet? More likely, the Astronomer Royal was not persuaded by Adams's evidence.

Today someone in Airy's position, like Brian Marsden and Daniel Green of the International Astronomical Union's Central Bureau for Astronomical Telegrams, receives several claims of a discovery of one sort or another each week, and they decide which ones are worth following. Had they encountered someone like Adams, obviously a serious student of the subject, particularly with Challis's support, they probably would take the claim seriously.

URBAIN JEAN JOSEPH LEVERRIER

In the same week that Airy wrote his letter to Adams, Leverrier entered the story. A student who twelve years earlier had graduated France's École Polytechnique,

Leverrier first worked as a chemist under the famous Joseph Gay-Lussac. In 1836 Leverrier was transferred to one of the provinces, but rather than accept what he thought was a demotion, he resigned from the government. In 1837, Gay-Lussac recommended him as an astronomy assistant at École Polytechnique. It was here that Leverrier became interested in the calculation of orbits.

Leverrier excelled at this. After presenting two remarkable papers to the Academy of Sciences, he decided to tackle another planet with strange perturbations—Mercury. Years later, Leverrier would propose that Mercury's strange orbit was the result of a gravitational pull from a planet, which he dubbed Vulcan, which was even closer to the Sun than Mercury. There is no Vulcan, we now know; and Mercury's orbital changes were finally explained by Einstein's Theory of Relativity.

Like most celestial mechanicians, Leverrier also tackled the difficult orbital problems of comets, particularly Messier's famous one of 1770. Then François Arago, one of France's best known astronomers and a member of the French Academy, suggested that Leverrier turn his considerable talent to the Uranus problem. It was the summer of 1845, and Leverrier spent months working on it.

THE SEARCH BEGINS

Completely unaware that Adams had done the same thing, Leverrier produced a paper that Airy saw and with which he was impressed. When the French observer sent Airy actual predictions for a new planet, Airy realized that Leverrier's predicted position, derived from different mathematics, was within one degree, or twice the apparent diameter of the Moon, from that of Adams. For some reason Airy never told Adams about the French calculations; nor did he inform Leverrier of Adams's work.

On June 29, 1846, John Herschel (William's son) and Challis met with Airy at Greenwich Observatory to discuss these developments. By early summer, Challis began a slow search across a large strip of sky, an inefficient procedure that Airy had suggested to him. Adams provided Challis with an ephemeris of possible positions of the new planet, with one position each twenty days from the end of July to the end of October. Why did Challis not use these positions to go directly to the predicted position? Instead he persisted in his slow, wide-area search. Challis later wrote that since it was such a new thing to go from a set of figures on a piece of paper to discover a planet in the sky, he did not consider that success was likely. Still, it would have been so much easier to dispose of the problem by quickly searching near Adams's positions.

Impatient and still hoping to get more action on his behalf, Adams decided to attend the meeting of the British Association for the Advancement of Science on

September 15 in order to make a presentation on the progress toward finding a trans-Uranian planet. However, when the hapless young Adams arrived he found that the relevant part of the meeting had taken place on September *14*, not 15—he was too late. Back at his observatory, Challis kept on looking, skipping over the new planet once or twice, and reporting no new results.

Leverrier ran into the same problem when he tried to mount his own search at the Paris Observatory. Although he had very specific positions near the star Delta Capricorni, he could not persuade anyone to look. Capricornus was now at its best observing position; in another few months it would be setting too early. Realizing that a search must start soon to have a good possibility of success, a desperate Leverrier asked a favor of Johann Galle at Germany's Berlin Observatory: "I would like to find a persistent observer who would be willing to devote some time to an examination of a part of the sky in which there may be a planet to discover. . . . You will see, Sir, I demonstrate that it is impossible to satisfy the observations of Uranus without introducing the action of a new Planet, thus far unknown; and remarkably, there is only one single position in the ecliptic where the perturbing planet can be located."[8]

SEPTEMBER 23, 1846

Galle discussed Leverrier's letter with the observatory director, none other than Johann Encke. The great celestial mechanician was skeptical, but curious, so he did allow his enthusiastic associate Johann Galle to begin an immediate search. On the night of September 23, 1846, Galle and a young student named Heinrich d'Arrest, who would later distinguish himself with the discovery of a comet, pointed the telescope immediately to the predicted position. At first they found only the stars that had been plotted on their charts. As they moved the telescope slowly in the direction of the ecliptic, where the planet would most likely lie, Galle called out positions of star after star while d'Arrest checked each one off on the chart. Galle finally called out the position of an eighth magnitude star, which d'Arrest could not identify. With mounting excitement both men looked at the star, then at each other, then rushed to inform Encke. Observing at a higher magnification, Encke even thought that the planet was not a point of light in the sky, but actually a round disk.[9] The planet was less than a degree from Leverrier's position, and less than a degree and a half from that proposed by Adams. "The planet whose position you have pointed out," Galle excitedly wrote Leverrier, *actually exists.*"[10]

THE POLITICAL STORM

Back in England, on September 29 Challis had somehow gone over the new planet twice without detecting it. Only four days after beginning his search, Challis's telescope came across Neptune. Incredibly, he noticed a star that seemed to have a small size to it, rather than being just a point of light. For some reason he didn't increase the magnification to confirm his suspicion, nor did he recheck the object the following night, even though it was still clear, because, he explained, bright moonlight prevented efficient observing.

On October 3, 1846, Sir John Herschel set out in the London *Athenaeum* the details of Adams's contribution to the discovery. Two days later, Challis penned an extraordinary letter to François Arago in France. In it he explained that he had been searching for a new planet as part of his star mapping work. Preposterously, Challis made no mention of Adams's extensive work in this letter.[11] On October 12, Challis wrote a very different letter to Airy, noting that he had detected the planet on August 4 and 12, but did not identify it as such at the time. He added that he strictly followed Leverrier's suggestions, with no mention of Adams. On the October 14 Airy, realizing the potential for a major controversy, wrote a diplomatic and congratulatory letter to Leverrier. In this letter he mentioned that English efforts had been made to predict the planet, pointing out that in his judgment they had not been as thorough as Leverrier's.

Airy's attempt at appeasement didn't work. Two days later Leverrier replied to Airy, demanding why, if he knew of other efforts, the Astronomer Royal of England hadn't bothered to even hint of them in his correspondence the previous summer. Then on October 17, Challis flip-flopped from his earlier report, publishing all his observations, as well as Adams's calculations and predictions, this time without mentioning anything about Leverrier.[12]

The die was cast; Challis's remarks enraged the French scientists. Arago in paticular was infuriated; he singled out Challis and Airy for their apparent attempt to take credit away from Leverrier. He took potshots at Adams as well, accusing the young man of trying to claim credit for a discovery for which he had no rightful claim. The French press, naturally, picked up the controversy and made it a major story, heaping criticism upon all the English astronomers, especially Adams, who, despite all his work, tried his best to stay out of the fray.

The storm raged for several weeks, and on November 13 Airy presented his full account of the Neptune story to a special meeting of the Royal Astronomical Society in London. Challis then described his search efforts, and finally Adams, in his first public appearance since the discovery, reported on his own research. The members of the society severely criticized Airy and Challis for their conduct during the affair, but applauded the effort and persistence of Adams, who gave a strict chronology of what happened when. "I mention these dates," he explained, "to show that my results were arrived at independently, and previously to the publication of those of M. Leverrier, and not with the intention of interfering with his just claims to the honours of the discovery."[13]

For a while it certainly appeared as though history would be unkind to Adams. The Royal Astronomical Society's Gold Medal, which had been previously presented to discoverers like Caroline Herschel, was not presented at all in 1846, but the Royal Society's Copley Medal, which went to William Herschel in 1782, was presented solely to Leverrier.

What should have been a joyful discovery experience for both Adams and Leverrier was marred by the other players, especially those in England. However, when the two discoverers finally met in June 1847—by then the recriminations had died down—they became lifelong friends until the elder Leverrier died in 1877. Adams continued his career, eventually joining the Royal Society himself and, years later, presenting the Gold Medal to Leverrier for his work on Mercury. Today Leverrier is honored in France with a large statue next to the Paris Observatory, a statue that reminds visitors of a discovery made over 150 years ago, and of a battle that drew attention from the fact that together, two young astronomers had pointed the way to a giant new planet circling more than 2 billion miles from the Sun.

Even as Neptune announced itself in the constellation of Capricornus, not far to the west in the sky lay the swirls of stars and clouds of the Milky Way, the densest part of our galaxy. Far to the south of Capricornus, the neighbor galaxies of the Magellanic clouds shone dimly for Southern Hemisphere observers. As astronomers began pointing their telescopes to these areas, it was time for humanity to spread its wings, and soar out beyond the solar system to these distant places.

ELEVEN

1912: HENRIETTA LEAVITT, HARLOW SHAPLEY, AND OUR PLACE IN THE MILKY WAY GALAXY

What a variable star "fiend" Miss Leavitt is—One can't keep up with the roll of the new discoveries.

—Charles Augustus Young,
Princeton University, March 1, 1905[1]

Miss Leavitt was one of the most important women ever to touch astronomy.

—Harlow Shapley, 1969

This is the story of how one woman made a discovery that led to the biggest change in our understanding of where we are in the universe since Galileo first pointed a telescope to the sky in 1610. The story really begins in the century before Galileo, when Ferdinand Magellan saw two faint clouds in the southern sky during his sixteenth-century travels. The explorer could have no idea of the important role they would play in humanity's understanding of the universe. Nor could Sir John Herschel when, during his observing voyage to South Africa during the 1830s, he made the first detailed observations of the stars in those clouds. John Herschel did take our understanding a step forward, however, through his correct belief that the clouds were systems of stars far beyond our own. It was left to a forty-year-old woman, Henrietta Swan Leavitt, to focus the attention of astronomers on just how important these galaxies were. Leavitt made a connection between the brightness of certain types of variable stars and the length of time it takes for them to complete a cycle of variation. Harlow Shapley then used that connection as a tool to find how far away those stars are, and hence to understand how large our galaxy is and where we are inside it. (Almost a century later, the Hubble Space Telescope can now detect variable stars farther and farther out in space. Thus, this yardstick is still used to determine accurate distances to more remote galaxies.)

A minister's daughter, Henrietta Leavitt was born in Lancaster, Massachusetts, on the fourth of July three years after the end of the Civil War. After high school in Cambridge she was accepted to Oberlin College, from which she transferred in 1888 to a place called the Society for Collegiate Instruction of Women, an institution later to become Radcliffe College. Although she was always interested in science, during her senior year there, in 1892, she had her first exposure to astronomy and loved it. After graduation, she became ill and was forced to reduce her activities for the next few years. When she was able to, she tried her hand at teaching, a difficult task since her illness had left her virtually deaf.

Unable to continue teaching, Leavitt volunteered for service at the Harvard College Observatory in 1895. Deafness was not a problem at Harvard; her associate Annie Jump Cannon also was very hard of hearing. Harvard College Observatory director Edward Charles Pickering assigned Leavitt to measure the brightness of stars near the pole. At the time, "Pickering's Girls" were hired for specific tasks and not encouraged to follow their own dreams. Within a year, however, Pickering would have high praise for her: "An interesting investigation has been made by Miss H. S. Leavitt on the photographic brightness of circumpolar stars."[2]

In 1900, Leavitt left the observatory. After remaining with her family, then in Wisconsin, for two years, she thought of returning to the observatory, even to continue working as a volunteer. "I am more sorry than I can tell you," she wrote Pickering, "that the work I undertook with such delight, and carried to a certain point, with such pleasure, should be left uncompleted." Pickering wrote back enthusiastically. He encouraged her to return to Harvard at once, at his expense; he offered her a salary, and even added that if she had to rejoin her family once more, she could take with her all the observatory materials, including photographs, to carry on her work.[3]

In August 1902 Leavitt returned to Harvard, where she stayed for the rest of her career. Her starting salary was thirty cents per hour, and her new assignment was to explore the magnitudes of variable stars taken from Harvard's Bruce Telescope at its southern station at Arequipa, Peru. A variable star is simply a star that is not constant in brightness. Depending on a star's type, its period of variation could be as short as a few hours or as long as several years in rare cases.[4]

It was while studying these southern plates, which included the Magellanic Clouds, that she became interested in the variable stars there. Leavitt's ability to discover variable stars on photographic plates was soon legendary, and Pickering later promoted her to head of the Department of Photographic Photometry.

E. C. Pickering has been accused of taking advantage of his women assistants. It is true that he could obtain their services for far less money than he could for men. It is also true, as Harvard astronomer Cecilia Payne-Gaposchkin later wrote, that Harvard's programs were essentially a one-man operation. Although "Miss Leavitt, Miss Canon and Miss [Antonia] Maury had emerged as independent scientists, their efforts had been channeled according to the plans of the Director."[5] Pickering took advantage of a way of life that was common at the time. The women he hired were the most intelligent and motivated that he could find, and he encouraged them to be the careful and productive scientists they were destined to become. He also went to bat for them: When his colleague Henry Norris Russell planned a collection of astronomical essays, Pickering was very upset that of the authors Russell had selected, the only Harvard astronomer chosen was Pickering himself. Where was Annie Cannon, he asked, Henrietta Leavitt, or Solon Bailey?[6]

CEPHEIDS

We now go back more than a century, to October 1784, when a teenage English deaf-mute named John Goodricke discovered that star Delta Cephei was changing in brightness. By the age of nineteen, Goodricke had discovered the variations of three stars, but Delta Cephei, it later turned out, was special. Its pulses were not caused by the periodic eclipsing of the star by a fainter companion. Instead, this star by itself varied periodically, and precisely, over a period of 5.37 days. It changes brightness as it expands and contracts on a regular basis; as it expands it fades, and as it contracts it brightens. Cepheid variables, precise as clockwork, were found to exist all over our galaxy. Harvard's Solon Bailey had noted large numbers of them in the globular clusters. After 1902, it was Leavitt's turn, and despite interruptions due to illness, she found them in the Magellanic Clouds as she continued her work.

To find these variable stars Leavitt used the method of "superposition." She would place a negative from a photographic plate taken on one date on top of a positive print of a photograph taken on another date. The black and white images should coincide, cancelling each other out. If they did not, Leavitt suspected the star was variable and would then repeat the process with a different set of photographs to confirm her discovery.[7] In 1904 she published her first lists of variable stars discovered in the Magellanic Clouds using this procedure. They included 152 variable stars in the Large Magellanic Cloud (LMC) and 59 in the Small Magellanic Cloud (SMC). Concentrating on the small cloud in the next year, she found an additional 843 variables there. These discoveries were published by the observatory as they occurred, and Pickering received several letters that year praising Leavitt for her enthusiasm and ability to discover variable after variable. One of these letters even suggested that Miss Leavitt be nominated for the Nobel Prize.

In 1908 Leavitt published an exhaustive list of 1,777 variable stars. From this huge list she focused on 16 Cepheid-type variables whose periods ranged in time from a short 1.25 days to 127 days. She also reported, somewhat tentatively at the time, that "it is worthy of notice that the brighter variables have the longer periods."[8] Although she had no idea at the time, Henrietta Swan Leavitt was onto one of the fundamental discoveries in astronomical history.

Leavitt's 1908 paper attracted almost no attention, however, possibly because it didn't list the variables according to magnitude or period, but according to the order of discovery. After all, Leavitt was primarily an observer of the sky on photographic plates. After the publication of that paper, Leavitt became seriously ill again and returned to Wisconsin. Honoring his earlier promise to her, Pickering sent her photographs by mail so she could continue her work.

When she returned to Harvard, Leavitt completed this phase of her work on the Cepheid variables in the Small Magellanic Cloud. She added nine additional Cepheids, enough to give her an understanding of the potential importance of her work. "Since the variables are probably at nearly the same distance from the Earth," she wrote, "their periods are apparently associated with their actual emission of light, as determined by their mass, density, and surface brightness."[9] In this paper Leavitt restated her earlier conclusion, this time with far more confidence:

> A remarkable relation between the brightness of these variables and the length of their periods will be noticed. In *Harvard Annals 60*, No. 4, attention was called to the fact that the brighter variables have the longer periods, but at that time it was felt that the number was too small to warrant the drawing of general conclusions. The periods of 8 additional variables which have been determined since that time, however, conform to the same law.[10]

Leavitt continued her work at the observatory. During the summer of 1912, she began her "Record of Progress" in which she listed all manner of research work. For the astronomer Henry Norris Russell she studied eclipsing variables; she measured luminosities of particular stars. She also tried to determine the spectra of faint stars. Pickering wanted her to return to her earlier work and determine the precise magnitudes of stars near the North Pole, stars that could be observed every night of the year and with which she had been working since she came to Harvard. This work was the start of an internationally agreed system of determining magnitudes of stars. However, it took Leavitt away from variable stars just at the high point of her work with them. Why Pickering did this is a mystery. It is possible that he was, at the time, not totally aware of the importance of Leavitt's work with the variable stars. In any event, the North Polar Sequence, Cecilia Payne-Gaposchkin would write years later, "ruthlessly relegated Miss Leavitt to the drudgery of fundamental photometry when her real interest lay in the variable stars that she had begun to discover in the Magellanic Clouds. She was the ablest of all the workers at Harvard at the turn of the century, but Pickering was a dictator, and his word was law."[11] Pickering insisted that Leavitt come up with a procedure to take magnitudes of stars determined with Harvard's several types of cameras, which ranged from the twenty-four-inch telescope in Peru to a one-inch telescope, and be able to reduce them all to the same photometric system.

"I cannot believe that he made so unrealistic a request," Payne-Gaposchkin wrote later. "It condemned a brilliant scientist to uncongenial work, and probably set back the study of variable stars by several decades."[12]

HARLOW SHAPLEY

A year after Leavitt's 1912 paper, a young astronomy student named Harlow Shapley received his Ph.D. from Princeton, traveled to Kansas City to get married, and set out for the Mt. Wilson Observatory in California. Mt. Wilson was at the astronomical frontier. It had a mighty 60-inch telescope, and an even mightier 100-inch was being planned. Years later, Shapley's colleague Bart Bok would imagine the two newlyweds, sitting on the front seat of a covered wagon, traveling across the land to conquer the new frontier. In fact, they traveled somewhat less dramatically by the Santa Fe Super Chief train, studying the behavior of variable stars most of the time.[13]

Shapley observed through the sixty-inch Mt. Wilson reflector, one of the finest optical telescopes ever built. He discovered RR Lyrae-type variable stars in globular clusters, and that those stars showed the very same period-luminosity relationship that Leavitt had discovered. By this time he had both confirmed Leavitt's work and expanded it to include closer globular clusters. He saw the enormous potential of Leavitt's discovery: If the period of variation gave a clue to the star's real brightness, then by comparing the absolute with the apparent brightness, we could calculate the distance to the star.

Shapley proposed that if the absolute magnitudes of the Cepheid stars in the Small Magellanic Cloud could be determined, then we would know the distance to this cloud. (A star's absolute magnitude is its brightness regardless of its distance from us; its apparent magnitude is its brightness as we see it, a factor influenced both by the star's absolute magnitude and its distance.) The average apparent magnitudes in one cluster would be brighter or fainter than in another cluster; because of this, Shapley was able to arrange the clusters by their relative distances from us. However to know how far these clusters were in light years,

Shapley needed to work the period-luminosity relation backwards. He needed to determine the absolute magnitudes to some nearby Cepheids, even just a few, by a different process. Why could he not use the period-luminosity relation with close stars? Although these stars are all closer than those in the Small Magellanic Cloud, they are at different distances from us. More important, since we are viewing them from within our own galaxy, these stars can appear brighter or fainter due to the thickness of interstellar dust that lies between each variable and Earth. A star in Cepheus, for example, might appear the same brightness as a star in Scorpius, but one might be dimmed more by interstellar dust.

The period-luminosity relation is hard to apply within our Milky Way Galaxy because the intervening dust prevents us from knowing the star's true luminosity. By studying the variable stars in the Small Magellanic Cloud, Leavitt avoided this problem. *All* of the Cepheids in the Small Magellanic Cloud are about the same distance from us—the distance of that galaxy. They are all affected equally by whatever interstellar dust lies in our own galaxy on the line of sight from them to us. The Small Cloud was chosen because it is a small, compact system, and all of it can be captured on a single photographic plate taken with a wide-field camera.

The key to using the period-luminosity relation to determine distances was first to determine the absolute magnitude of nearby Cepheid variables though by using a different process: In 1918, Shapley used the Hertzsprung-Russell diagram that plots star brightness against surface temperature, to determine the absolute magnitudes of eleven nearby Cepheid variable stars. He was able to calculate the surface temperature by observing each star's spectrum and fitting it onto the diagram. This method works for stars bright and close enough to reveal their spectra to a telescope. Stars in the distant globular clusters and the Magallanic Clouds were too faint to reveal their spectra.

Once Shapley had the absolute magnitudes to the eleven Cepheids in our own galaxy, he had the key to their distances and the calibration he needed to determine distances to the globular clusters and the Magellanic Clouds. He then concluded that the globular clusters were very far—as much as 50,000 light years—away, and that the Small Magellanic Cloud was more than three times that distance. Shapley went on to place the center of the Milky Way some 50,000 light years away.

Until this time, astronomers were ready to accept that our Sun was at the center of the galaxy. Thanks to this work by Leavitt and Shapley, all that changed. The Sun was far from the galactic center, actually in its outskirts. "I stayed with the Cepheids and clusters during those early years at Mount Wilson," he wrote, "until I crashed through on the distances and outlined the structure of the universe. . . . I plotted the clusters and looked at what I had. Finally I hit upon using the period-luminosity relation that had been foreshadowed by Miss Henrietta Leavitt at Harvard in a paper published in 1912. Her paper dealt with only twenty-five stars and did not deal with their distances at all. So I went after the distances, and that was helped by Ejnar Hertzsprung's work."[14]

SHAPLEY MEETS LEAVITT

In April 1921, at the young age of twenty-six, Shapley arrived at Harvard as director of the observatory. Only nine years had passed since the landmark paper that would make Leavitt one of the most famous astronomers, but in this time she had become once again ill, this time with cancer. Tragically, Leavitt was already near death at the young age of fifty-three. "Miss Leavitt was one of the most important women ever to touch astronomy," Shapley wrote in his own memoirs. "She had many handicaps, including deafness. . . . She worked about a year longer—mostly on the Magellanic Clouds, which became one of our major projects. I worked with her on these, but she was not able to do much by that time. One of the few decent things I have done was to call on her on her death bed; it made life so much different, friends said, that the director came to see her."

Leavitt's unfinished interests in variable stars would be continued by Ceclia Payne-Gaposchkin, whose work extended the variable stars in the Large and Small Magellanic Clouds to almost 4,000. "When I first came to Harvard," she wrote about her arrival two years after Leavitt's death, "the Observatory was at the parting of the ways. I never saw Pickering, never knew Miss Leavitt, though their shadows could still be discerned. I heard tell that Miss Leavitt's lamp was still to be seen burning in the night, that her spirit still haunted the [photographic] plate stacks. I suspect that some credulous soul (and there [were] such in those days) had seen me from afar, burning the midnight oil. Shapley had given me the desk at which she used to work."[15]

TWELVE

1919:
EDDINGTON, EINSTEIN,
MERCURY, AND AN ECLIPSE

*This discovery was, I believe, by far the strongest emotional experience in
Einstein's scientific life, perhaps in all his life. Nature had spoken to him.*
—Abraham Pais, on Einstein's discovery that
the perihelion precession of Mercury's orbit
could be explained through General Relativity[1]

At the same time that Shapley was making us aware of the size of our galaxy, other scientists were exploring the forces that make it work. This part of our story begins in the 1890s, when a teenager named Albert Einstein asked himself what the world would look like if he rode on a beam of light. The world would look frozen in time, its clocks motionless as in a photograph. A decade later, Einstein was unable to find a job in physics, but he did find one at the Swiss patent office, where in between looking over patent applications, he had time to ponder questions such as the relationship between matter and energy.

Einstein published his Special Theory of Relativity in a 1905 article, and later that year, Einstein attached an additional thought, the equation $E=mc^2$. In those simple letters lay the phenomenal idea that mass and energy are equivalent.[2] Late in 1915, Einstein's General Theory of Relativity offered a new definition of gravitation that related it to space and time. Relativity was now applicable not just to specific situations but throughout physics. In Einstein's physics, unlike that of Isaac Newton, gravity is not a force but geometry. As any object moves, whether it is a baseball, a planet, or a star, it follows a non-Euclidian geometric path that is shaped by the unified effect of mass and energy. In almost every physical possibility, Newton's laws, conceived during the time of Huygens and Halley, do work, or at least they work at the level to which we can measure them. But where an enormous amount of matter is involved, as in the space near a star or a black hole, these laws do not work.

After 1905, Einstein's Special Theory of Relativity aroused the interest of physicists, but since it was generally thought to be unprovable, it was predominantly seen as interesting but something that would never replace Newton. The General Theory of Relativity was of even greater interest to physicists, though some connection to reality needed to be established. Einstein found that connection through the nagging problem of the orbit of Mercury, whose perihelion, or closest point to the Sun, precesses, or shifts, by forty-three seconds of arc per century beyond what Newton's law predicts.

We have already seen how Leverrier, who discovered the anomaly, attempted to explain this orbital change through the gravitational pull of a planet even closer to the Sun called Vulcan. On November 18, 1915, Einstein wrote that his General Theory of Relativity solves the problem of "the secular rotation of the orbit of Mercury, discovered by Le Verrier, . . . without the need of any special hypothesis." As his friend and colleague Abraham Pais wrote,

> This discovery was, I believe, by far the strongest emotional experience in Einstein's scientific life, perhaps in all his life. Nature had spoken to him. He had to be right. "For a few days, I was beside myself with joyous excitement." Later, he told [Adriaan] Fokker that his discovery had given him palpitations of the heart. What he told [Wander] de Haas is even more profoundly significant: when he saw that his calculations agreed with the unexplained astronomical observations, he had the feeling that something actually snapped in him.[3]

THE ECLIPSE OF MAY 29, 1919

Demonstrating something mathematically was something at which Einstein excelled, but doing it observationally required the patience of another kind of person. The English astronomer Arthur Stanley Eddington was that person. Eddington never really knew his educator father, who died in 1884 when Arthur was only two. Before his fourth birthday Eddington became interested in large numbers—he decided to count all the letters in the Bible, and although he didn't finish this prodigious task, he did finish Genesis, and also learned the 24×24 multiplication table. Even more telling was that he enjoyed being taken outdoors on clear evenings so he could count the stars.[4]

At the age of fourteen Eddington received the school honor of first class with distinction in mathematics. In the fall of 1902, he entered Trinity College at Cambridge University, where his classmates perceived him as quiet but friendly, and where he enjoyed playing chess. His three years at Trinity, first as a student, then as an instructor, ended in 1906, when he recorded this entry in his diary: "Jan 15: I received a letter from the Astronomer Royal offering me his nomination to the post of Chief Assistant at the Royal Observatory."[5]

Eddington stayed at Greenwich until 1913. Those years were good ones, punctuated by his love of cycling, chess, the stars, and in a precursor to his later work, his first attempt to witness a total eclipse of the Sun. Unfortunately, on October 10, 1912, the eclipse site he traveled to in Brazil was completely rained out. It was time to move on, back to Cambridge Observatory to study the two fields in which he would leave his mark on the sands of time, relativity and astrophysics.

PROVING RELATIVITY

Eddington was about to find that the rained-out eclipse was a harbinger of his future research. He knew that although Einstein was delighted that Mercury offered some observational evidence that his theory was correct, there needed to be a more dramatic line of evidence. Eddington thought that a solar eclipse might provide that evidence.

Einstein's theory, Eddington noted, "leads to interesting conclusions with regard to the deflection of light by a gravitational field," and that it could be tested through an experiment.[6] Eddington needed to answer first the question of whether light has weight, as Newton suggests, and if the answer was yes, is the amount of deflection, or bending of light, in agreement with Newton, or with Einstein? By photographing a star near the Sun, and then comparing its position with that on other photographs taken when the star is far from the Sun, Einstein's relation could be tested on the deflection of that star. The deflection of the light of a star as it passes close to a large gravitational mass like the Sun is in the opposite direction from the Sun; the star will appear to be displaced outward, or away from the Sun, by the same amount as the deflection of its light toward the Sun. Although this proposed test is fine in theory, when the star's light passes that close to the Sun, it is so lost in the glare of the Sun that we cannot observe it. There is just one time that such an observation is possible, and that is during a total eclipse of the Sun.

The bending of a distant star's light is detectable only with stars when they are near our line-of-sight to the Sun. These stars can be seen only during a total eclipse. The problem is that even during an eclipse all is not dark. The Sun's corona is bright enough to blot out the light from faint stars normally visible with ease through a telescope. The only eclipse that would work is one in which the Sun is near a cluster of moderately bright stars. By some trick of providence, just such a rare coincidence was about to happen for Eddington. On May 29, 1919, the Sun would be eclipsed while it was close to the Hyades star cluster in Taurus. The cluster would be just south of the eclipsed Sun, its stars bright enough to be captured on the photographic films of the time.

Eddington wrote,

> In a superstitious age, a natural philosopher wishing to perform an important experiment would consult an astrologer to ascertain an auspicious moment for the trial. With better reason, an astronomer to-day consulting the stars would announce that the most favorable day of the year for weighing light is May 29. The reason is that the sun in its annual journey round the ecliptic goes through fields of stars of varying richness, but on May 29 it is in the midst of a quite exceptional patch of bright stars—part of the Hyades—by far the best star-field encountered. Now if this problem had been put forward at some other period of history, it might have been necessary to wait some thousands of years for a total eclipse of the sun to happen on the lucky date. But by strange good fortune an eclipse did happen on May 29, 1919. Owing to the curious sequence of eclipses a similar opportunity will recur in 1938; we are in the midst of the most favorable cycle. It is not suggested that it is impossible to make the test at other eclipses; but the work will necessarily be more difficult.[7]

The "curious sequence" refers to the Metonic cycle, in which eclipses on a particular date might be followed by a second eclipse on the same date, nineteen years later.

Eddington's plan was to measure the displacement of a star in comparison with stars that are farther from the Sun and not displaced. The next challenge was to plan the expedition, no small task near the end of World War I. Getting official blessings, and funding, for two expeditions to Africa and to Brazil to observe an eclipse in order to confirm the ideas of a German scientist seemed a tall order, but Sir Frank Dyson, England's then Astronomer Royal and a very good friend of Eddington's, persuaded the government that this eclipse presented a unique opportunity. He managed to shake £1,000 loose from the treasury for the expedition.

The winter before the eclipse, when the Hyades cluster was high in the nighttime sky, Eddington photographed the cluster using the same telescope that would be brought to Africa. Showing no deflections, this photograph would serve as a base for comparison. The night before the expedition set sail, there was a discussion about just how much deflection the star would suffer: If it were but 0.87 arcseconds for stars at the edge of the Sun, then it would confirm Newton's classical theory of gravitation; if it were 1.75 arcseconds, Einstein would triumph. The evening before sailing, Edwin Cottingham, who was to accompany Eddington, was chatting with Eddington and Dyson. What would happen, he asked, if the star's deflection was double what Einstein had predicted? "Then Eddington will go mad," Dyson laughed, "and you will have to come home alone!"

Eddington described this fateful expedition in his notebook:

> We sailed early in March to Lisbon. At Frunchal we saw [the other expedition] off to Brazil on March 16, but we had to remain until April 9 . . . and got our first sight of Principe in the morning of April 23. . . . [A]bout May 16 we had no difficulty in getting the check photographs on three different nights. I had a great deal of work measuring these.

The group arrived in Principe with a thirteen-inch-diameter, eleven-and-one-third-foot-long refractor, which they "stopped down" (reduced in diameter using an iris) to eight inches to improve the image quality. The telescope was mounted in a fixed position, and a coelostat mirror directed the light from the stars into the telescope.

> On May 29 a tremendous rainstorm came on [Eddington wrote]. The rain stopped about noon and about 1:30 when the partial phase was well advanced, we began to get a glimpse of the sun. We had to carry out our programme of photographs in faith. I did not see the eclipse, being too busy changing plates, except for one glance to make sure it had begun and another half-way through to see how much cloud there was. We took 16 photographs. They are all good of the sun, showing a very remarkable prominence; but the cloud has interfered with the star images. The last six photographs show a few images which I hope will give us what we need.[8]

A year later Eddington expanded on the details of those few minutes of total eclipse:

> There was nothing for it but to carry out the arranged program and hope for the best. One observer was kept occupied changing the plates in rapid succession, whilst the other [presumably Eddington himself] gave the exposures of the required length with a screen held in front of the object glass to avoid shaking the telescope in any way.
>
> > For in and out, above, about, below,
> > 'Tis nothing but a Magic Shadow-show
> > Played in a box whose candle is the Sun
> > Round which we Phantom Figures come and go.
>
> Our shadow-box takes up all our attention. There is a marvelous spectacle above, and, as the photographs afterwards revealed, a wonderful prominence-flame is poised a hundred thousand miles above the surface of the sun. We have no time to snatch a glance at it. We are conscious only of the weird half-light of

the landscape and the hush of nature, broken by the calls of the observers, and beat of the metronome ticking out the 302 seconds of totality.[9]

Eddington and Cottingham raced through sixteen exposures, ranging in exposure time from two to twenty seconds. The first ones beautifully captured the prominence, but no stars. Toward the end of totality, the clouds parted, and one photographic plate recorded five members of the Hyades cluster. Once the pictures had been processed, Eddington made his first measurements at the eclipse site. The two photos were placed "film to film" in the measuring machine so that the star images were close to identical. "In comparing two plates," Eddington wrote, "various allowances had to be made for refraction, aberration [of the telescope lens], plate-orientation, etc."[10]

"June 3," Eddington wrote of that critical day in his *Notebook*, "We developed the photographs, 2 each night for 6 nights after the eclipse, and I spent the whole day measuring. The cloudy weather [during the eclipse] upset my plans and I had to treat the measures in a different way from what I intended, consequently I have not been able to make any preliminary announcement of the result. But the one plate that I measured gave a result agreeing with Einstein."[11]

As he completed the reduction of this plate, Eddington realized the significance of his result. Turning to his colleague, he smiled and said, "Cottingham, you won't have to go home alone." They packed their priceless plates and returned to England. Four additional plates of a more sensitive type which could not be developed in the hot African climate were developed at Cambridge. With trembling hands Eddington examined these plates. The clearest one, he noted at last, confirmed the result shown on the first successful plate.

As Carl Sagan would say decades later, extraordinary claims require extraordinary proof. For a result as crucial as this one, Eddington had to ensure that instrument errors could not have led to the result. As a check, Eddington photographed a different star field, at night, with his arrangement at Principe. If the "Einstein deflection" were the result of a telescope error of some kind, it would have turned up in these check plates. No changes were found in those stars.

What of the expedition that went to Brazil? That team viewed the eclipse earlier in the day, and had better luck. The weather on eclipse day was superb, so the members remained at their site for two additional months in order to photograph the same region of sky under cover of morning darkness. They had two telescopes, one similar to the African scope, and a much longer, four-inch-diameter, nineteen-and-one-third-foot-long refractor. But the team had its problems as well. When the Brazil expedition returned home finally, its measured deflections with the first telescope did *not* agree with Einstein, but with Newton! This was a shocking result, one which would inevitably delay any announcement, one way or another, about relativity. Eddington suspected that the Sun's rays in the clear

Brazil sky might have distorted the coelostat—the flat mirror that reflected sunlight into the telescope—thus distorting the images. The cloudy weather in Africa seemed to have been a blessing. "At Principe," he wrote, "there could be no evil effects from the sun's rays on the mirror, for the sun had withdrawn . . . shyly behind the veil of cloud."[12]

Were the Brazil results distorted by the Sun-facing mirror, or did they doom relativity? The final verdict had to wait until their plate-measuring device could be modified to accept the seven different-sized plates taken through the four-inch refractor. The images on these plates were perfect, as was the result they revealed: a deflection in strong agreement with the results in Africa!

Einstein was thrilled when he heard of this result. "Lieber Herr Eddington!" he began. "I should like to congratulate you on the success of this difficult expedition," he wrote. "I am amazed at the interest which my English colleagues have taken in the theory in spite of its difficulty. . . . If it were proved that this effect does not exist in nature, then the whole theory would have to be abandoned!"[13]

The announcement of these astonishing results put Einstein's Theory of Relativity onto the front pages of newspapers around the world. The man from the patent office had given humanity a new way of looking at gravity, and Eddington summed it up in verse based on his beloved poem translated by Fitzgerald as the *Rubáiyát of Omar Khayaam*:

Ah Moon of my Delight far on the wane,
The Moon of Heaven has reached the Node again
But clouds are massing in the gloomy sky
O'er this same Island, where we labored long—in vain?

And this I know; whether EINSTEIN is right
Or all his Theories are exploded quite,
One glimpse of stars amid the Darkness caught
Better than hours of toil by Candle-light

Ah friend! Could thou and I with LLOYDS insure
For Gold this sorry Coelostat so poor,
Would we not shatter it to bits—and for
The next eclipse a trustier Clock procure . . .

The Clock no question makes of Fasts or Slows,
But steadily and with a constant Rate it goes.
And Lo! The clouds are parting and the Sun
A crescent glimmering on the screen—It shows!—It shows!!

Five Minutes, not a moment left to waste,
Five Minutes, for the picture to be traced—
The Stars are shining, and coronal light
Streams from the Orb of Darkness—Oh make haste!

For in and out, above, about, below
'Tis nothing but a magic *Shadow* show
Played in a Box, whose Candle is the Sun
Round which we phantom figures come and go . . .

Oh leave the Wise our measures to collate
One thing at least is certain, LIGHT has WEIGHT
One thing is certain, and the rest debate
Light-rays, when near the Sun, DO NOT GO STRAIGHT.[14]

HOW STARS WORK

At the same time that Eddington was working on relativity, he was also trying to understand astrophysics, or the physics of the stars. In 1916 he derived the formulae that explain why a star continues to shine and normally does not blow up. What keeps it from blowing up is a process he called radiative equilibrium. For a tiny star much smaller than the Sun, like Proxima Centauri, radiation pressure does not have to be high at all compared to gravitation for the star to hold together. For stars like the Sun, the pressure has to be about 10 percent of gravitation to keep the star together. For the huge, short-lived stars like P Cygni, radiation pressure is as high as 80 percent that of the star's gravity. "A gas sphere under the influence of two opposed and nearly balanced forces would probably be on the verge of instability. This suggests that though the radiation pressure does not actually break up the mass [of the star], it produces a state in which the mass will, under ordinary circumstances, be broken up by rotation or other causes. Occasional large masses with exceptionally low angular momentum may survive, as in the case of Canopus."[15] Eddington believed that a star's own force of contraction was not enough to keep a star alive for more than 100,000 years. "We can only avoid a short time scale by supposing that the star has some unknown supply of energy." That energy, of course, is nuclear fusion, but that was not known until the 1930s.

Until 1923, Eddington's main work centered on relativity, but his discoveries and formulations in astrophysics came quickly afterward. In 1923 he derived a formula to relate the luminosity of a star to its mass, and in the same year correctly interpreted high-density, white dwarf stars as being formed of matter so dense that atomic electrons have collapsed from their orbits, a substance we now call degenerate matter. Eddington became as famous for his work on stellar physics as for relativity. There is a story about Eddington which is almost certainly apocryphal, but too good to omit. Flush with his new understanding about the constitution of the stars, but having not yet published his work, one clear evening he was sitting outdoors with a friend who remarked how beautiful the stars are. "Yes," Eddington replied, "Even more so now that I know how they work!"

Despite increasing pain late in his life, Eddington worked on his science, and on books written for the general public. On the night of November 6, 1944, the pain was in too consuming for him to continue. He died on November 22.

"We may look on the universe as a symphony played on seven primitive constants as music is played on the seven notes of a scale," Eddington told us.[16] The universal notes include constants like the masses of the electron and proton, the electron's charge, the velocity of light, and the Hubble parameter which we visit in the next chapter. What an artistic way for a scientist to look upon the universe he understood so well.

THIRTEEN

1924:
EDWIN HUBBLE

The Man Who Defined the Universe

> *On this plate (H335H) three stars were found, 2 of which were novae, and 1 proved to be a variable, later identified as a Cepheid—the first to be recognized in M31.*
>
> —Edwin Hubble, October 1923, Observing Log, p. 156[1]

Henrietta Leavitt's discovery of the relation between period and luminosity of the Cepheid variables led directly, thanks to Harlow Shapley, to a new understanding of our place in the Milky Way. This was all made possible by 1920, but the effects of Leavitt's work did not end there. With the greater power of the 100-inch telescope, which celebrated its "First Light" atop southern California's Mt. Wilson in 1918, Edwin Powell Hubble was able to find Cepheid variables in more remote galaxies. But to stretch the distance scale beyond the nearest galaxies, a different yardstick was needed. This chapter tells the story of how Edwin Hubble, astronomer extraordinare, used the Cepheid variable yardstick as the foundation for a new yardstick, the redshift, that would permit distance measurements to the very edge of the universe.

Although Hubble's story takes up where Shapley's left off, its origins go back to the turn of the century to a small, remote observatory located in the town of Flagstaff, Arizona. That observatory was run by Percival Lowell for the purpose of observing Mars and learning about the solar system. In 1903, Lowell became intrigued by the many fuzzy spots in the sky called spiral nebulae. Messier and the Herschels had catalogued them, and in 1850 the Earl of Rosse, using the largest telescope in the world at the time, sketched their cloudy swirls. Lowell shared the common belief that these nebulae might be newly hatching solar systems. He asked his chief observer, Vesto M. Slipher, to take very long exposure photographs of some of these nebulae with a spectroscope attached to the telescope. The resulting images might shed light on the nature of these nebulae.

They did, but not in the way Slipher or Lowell expected. After taking exposures as long as three nights (the telescope would be closed, then reopened, each night to resume the exposure), Slipher was puzzled at his results. The spectra, which looked like an irregular series of lines somewhat like a bar code reading, showed that while the "H and K" lines of calcium in the spectrum of the nebula in Andromeda were shifted toward the left, or blue end of the spectrum, the spectra of the fainter nebulae were shifted toward the red end.

Even then astronomers knew that light waves behaved like sound waves. When a train whistles or a driver honks the horn when passing, the pitch of the sound drops. The same effect, known as the Doppler shift, happens with light. The spectra seemed to be telling us that most of the nebulae were moving away from us.

THE GREAT DEBATE

In 1920, the Leavitt-Shapley discovery was still the only yardstick for measuring distances to far-off objects. Where it was leading, however, was still an open question when, on April 26, Harlow Shapley and Heber Curtis came together in Washington to discuss the state of the universe. The question was this: What were the spiral-shaped nebulae, and how far away were they? From his studies of the Cepheids, Shapley had already established that our galaxy was some ten times bigger than scientists had previously thought. Because our own galaxy was so big, Shapley reasoned that the spiral nebulae were related to our own galaxy, and could not lie very far outside it.

Heber Curtis took the side of the old school: Our galaxy was far smaller than Shapley's Cepheid measurements had indicated. Since our galaxy is small, those remote spiral-shaped fuzzy patches are comparable in size and nature to our own galaxy, and are probably far away from it. Curtis had strong support from an important scientist: Back in 1914, Arthur Eddington suggested that the remote nebulae were really galaxies like our own.

Although no one knew it at the time, history records that Curtis won the debate by using the wrong argument—against Shapley—to come up with the right answer. The spiral nebulae are comparable in size to the Milky Way, and they are at incredibly vast distances from it. Although Shapley was correct in his argument about the size of our galaxy, he was wrong about the nature and distance of the spiral nebulae. Curtis was right about the nature of the spirals, but for the wrong reason! Using the observational data of the time, both scientists did the best they could. It is interesting how scientific debates compare with political ones, where poise and the power of persuasion are paramount. In scientific debates neither poise nor persuasion have anything to do with it; Nature itself watches the arguments, quietly keeping score and choosing the winner. Just four years after the Shapley-Curtis debate, Nature did chime in with its verdict, and for that story, we move on to the life and career of Edwin Powell Hubble.

EDWIN HUBBLE: COSMIC MAN

Born in 1889, Edwin Hubble was a precocious child who, by age twelve, was already attending eighth grade, two grades ahead of his peers. In 1906 he entered the University of Chicago. Although he started there as a freshman, his sights were already set on the elite schools in England, particularly Oxford. He joined Kappa Sigma, one of the campus fraternities, despite its requirement which Hubble abhorred that all "pledges" attend dancing parties. Notwithstanding the inconvenience of dances, Hubble didn't want to let anything get in the way of his goal to become an astronomer. But unlike those of earlier scientists like Tycho and Galileo, Hubble's father did not favor his son's professional goal. More like Constantijn Huygens, John Powell Hubble wanted Edwin to be a lawyer. To appease his father, Edwin, as he took university courses in astronomy, tried to satisfy the pre-law requirements.

Edwin's behavior during his second year did little to assuage his father's concerns about his future. Kappa Sigma's house happened to be next to the residence for the university's theology students. Anxious to "raise hell" with them, or at least to tease them for having to wear black suits to class, Hubble and his friends waited until the time that the students' black suits were delivered, on schedule, from the dry cleaner. As the suits emerged from the truck, the Kappa Sigmans tossed eggs on them.

Although a prank like this would hardly have deserved mention in later years—as a university student I too took part in stunts as bad as that—at the University of Chicago, such stunts earned punishment from the highest levels of the administration. The gravest embarrassment of all was that the pranksters' parents were notified of the infraction. When Edwin next went home his father lectured him that his conduct might well have cost him the Rhodes Scholarship he craved. Edwin's grandfather disagreed. Rather enjoying the story, he thought Edwin deserved praise for ingenuity, not criticism.[2]

Back at school Edwin redoubled his efforts, and not just at his studies. Always an athlete, Hubble made the university's varsity basketball team, which enjoyed an almost undefeated season that brought them into the national championships. Chicago won the first game handily, but in the second game the Pennsylvania Quakers fought hard, very nearly beating Chicago. Late in the game, however, Chicago triumphed, edging the Quakers 17–16. Hubble did everything he could to please the Rhodes trustees, including running for vice-president of his senior class. The egg incident did not prevent him from winning the coveted Rhodes Scholarship. Coupled with his athletic prowess and high grades, Hubble was chosen as Rhodes Scholar from Illinois.

On September 7, 1910, a few months after the appearance of Halley's Comet, Hubble boarded the *Canada*, docked in Montreal, for his journey to Queen's College at Oxford. His father was still insisting that he study law, not astronomy, and the obedient son obliged: "Have decided to cut out all diversions and study nothing but law," he wrote his parents. "My ambitions are a book, an easy chair, and a fireplace."[3]

At the same time, however, Hubble was enjoying the company of Oxford's astronomy professor Herbert Hall Turner. He did reveal to his parents later that he was enjoying some diversion into astronomy, but even though he wanted more than anything to become an astronomer, he held his love and respect for his family higher. Concern for his father's failing health prevented him from trading his own ambitions for his father's plan. In January 1913, however, he received a cable from America with the news that his father had died. Soon after Hubble returned home with his Oxford cane, cape, and newly minted British accent intact, his three Oxford years over.

Hubble varied his course work as he ended his time at Oxford, including a class in literature. After he returned to the United States, he took a position as a teacher of Spanish and physics at a high school in Indiana. During that year he knew that the course of his future was now open to him; if he was to change direction and turn to his love of astronomy, now was the time. In May 1914 he wrote to his old astronomy professor at the University of Chicago, who suggested that he contact Edwin Frost, the director of the Yerkes Observatory that was a part of the University of Chicago. Two years earlier Frost had been appointed editor of the prestigious *Astrophysical Journal* (*Ap. J.*), who along with co-editors George E. Hale and Henry G. Gale were known as "the three meteorologists."[4] (The flagship publication of the American Astronomical Society, *Ap. J.* is still the journal of record for the most important papers in astrophysics.) Eager to admit the ambitious Hubble to his graduate program but unable to offer a scholarship to him, Frost answered Hubble's request with the offer of a tuition scholarship.[5] By his second year Hubble was well into his area of research; the investigation of the nebulae that V. M. Slipher had examined years earlier by means of photography.

HUBBLE'S VARIABLE NEBULA

Although in his early years at the observatory, Hubble did not often use the observatory's gem of a telescope, its forty-inch refractor, he did make good use of the smaller twenty-four-inch reflector there. It wasn't long before Hubble made his first discovery at Yerkes. He decided to try photographing a strange nebula, NGC 2261, in the constellation of Monoceros, that looked like a comet but which we now know to be a nebula that is lit by the young variable star R Monocerotis. The star's light both reflects off the nebula and shines through it, causing the nebula's atoms to ionize and reemit it; thus the nebula is both a reflection and an emission nebula. Hubble photographed it many times during the observing season that began in the fall of 1915. When he compared his pictures to images taken in 1908, he saw that the nebula's west side had expanded and had become more convex. Yerkes Observatory was an ideal place for making one's first discovery; besides Frost, the observatory was the professional home of Edward Emerson Barnard, one of the world's most successful comet discoverers. Taking advantage of Barnard's long observing experience, Hubble worked with the older astronomer to confirm his discovery.

WAR AND SPACE

In April 1917, the United States declared war on Germany, and Hubble immediately applied for the Officer Reserve Corps. His dissertation on the faint nebulae was complete, but Hubble was unhappy with it. Although neither he nor Frost believed it to be his best work, the pressure of war, and Hubble's desire to be part of it, led to his degree being granted on condition that Hubble complete his work later. The new Dr. Hubble quickly became a captain, then a major, and remained with the armed forces until the end of the summer of 1919.

Meanwhile, on November 1, 1917, the 100-inch at Mt. Wilson was pointed to the stars. That first night was a scary one. With a clear sky, Hale and his associate Walter Adams, accompanied by the famous poet Alfred Noyes, looked through the large telescope. According to Walter Adams's account, when they looked at Jupiter, they were stunned at the image. Instead of one Jupiter, several images of the planet squiggled over each other—the telescope's performance was awful.

Would the largest telescope in the world be a complete bust? The group in the dome couldn't understand why the telescope was performing so poorly. Someone did come up with an incredible thought, though: During the warm, sunny hours of the preceding afternoon, the dome had been left open as workers were making final adjustments. Could sunlight have fallen on or near the mirror? The group decided to give the mirror time to cool, and Hale and Adams returned several hours later in the predawn hour of 2:30 A.M. Again according to Adams, Jupiter had set, so with trembling hands they moved the telescope to Vega. Their first look confirmed their highest hopes: the single image of the star was dazzling and beautiful.

The telescope's main mirror, it seemed, was a work of art whose final polish was applied painstakingly by the optical expert George Ritchey. After that stressful night, the 100-inch looked forward to a grand career of uncovering secrets of the universe. But we cannot be certain of the details of Adams's story. Why, for example, would they have chosen Jupiter as an early evening object, when it was low in the east and subject to the poor conditions that always exist near the horizon? (In fact, Jupiter was almost exactly at the same position in the sky as it was when we wrote this chapter!) Moreover, there is no chance that the tired and nervous observers could have looked at Vega in the predawn hours, for Vega had long since set. It is possible that Adams simply had the objects reversed—Vega was high in the west that evening, and Jupiter, with the gibbous Moon nearby, was almost overhead by 3 A.M. Whatever happened that auspicious night, both telescope and observer needed the next two years to be prepared for its assault on the universe. Hubble was in the military; the telescope was undergoing tests.

In the fall of 1919, Hubble arrived at the newly opened 100-inch telescope to resume his career of peering to the stars.

A CEPHEID IN THE "ANDROMEDA NEBULA"

As a staff astronomer at Mt. Wilson, Hubble now had almost unlimited access to the 100-inch telescope, and his interest naturally turned to the faint nebulae that were the subject of his dissertation. On October 4, 1923 (coincidentally thirty-four years to the day before the launching of the first artificial satellite by the Soviet Union) Hubble took a forty-minute exposure of one of the spiral arms of the Andromeda Nebula, which at that time of year is well placed for viewing. Despite the poor observing conditions, Hubble thought that the plate contained a nova within that spiral arm. If true, then this would be strong evidence that the Andromeda spiral was a distant maelstrom of stars, and not something close to our own galaxy.

The following night Hubble took a second plate, this time under a much clearer sky. That plate, which Hubble listed in his log as H335H, confirmed the nova, and there seemed to be two additional ones. Hubble returned to his Pasadena office with his precious plates and set to work on the three new stars by studying older plates, some of which had been taken by Harlow Shapley. The evidence was now obvious to the delighted Hubble: One of the three interesting stars on H335H was a Cepheid variable with a period of thirty-one days. Excitedly noting the discovery and importance of the plate in his log, he then took out a pen and wrote "VAR!" in capital letters on the original plate.[6]

Hubble's variable nebula, with its changing size and shape, surrounded by stars whose proper motions could be measured, must be relatively nearby. The nebula in Andromeda was another story. Hubble was certain that it was another galaxy, an extragalactic nebula a large distance from the Milky Way. In February 1924 he wrote Shapley that he had found a Cepheid in the Andromeda Nebula, and that using Shapley's own scale, he had determined that the distance to the nebula must be at least a million light years. Shapley was uncertain that this single Cepheid, with its relatively long period, could be effectively used to calculate a distance, but he couldn't ignore the fact that his nemesis, Hubble, had just rejudged the old Curtis-Shapley debate, and that Curtis was the winner. Actually Shapley didn't really lose it: His Milky Way galaxy was as large as he had thought, but the universe, it appeared, was vast. Years later Walter Baade would use the 200-inch telescope at the Palomar Observatory to learn that the Hubble distances really needed to be doubled: we now think that the Andromeda system is some *two* million light years away.

Once Shapley understood that Hubble's nebulae were clearly galaxies like our own, he engaged in a new debate about what to call them. Hubble still referred to them as nebulae, or extragalactic nebulae, but Shapley insisted that as long as they were like our Milky Way they should be called galaxies. Shapley's own popular book, *Galaxies*, helped confirm the public impression that these distant objects were similar to our own galaxy.[7]

When I first entered the Montreal Observatory of the Royal Astronomical Society of Canada in 1960, Miss Isabel Williamson, the director of the society's observing program, instructed us that the correct name for these distant objects was *galaxy*, not *nebula*. In fact, if we were ever caught saying "Andromeda Nebula" instead of "Andromeda Galaxy," we were fined twenty-five cents! Shapley would have been proud.

THE REDSHIFT

On September 10, 1923, the Mt. Wilson astronomers were preparing to observe a total eclipse of the Sun. On that day the shadow of the Moon would swing near San Diego. The event was, surprising for that area and time of year, obscured by a freak thunderstorm, but after the eclipse Hubble met Grace Leib, the daughter of the banker who had funded one of the observing teams for the eclipse. The couple was married on February 26, 1924.

The two years following the wedding were busy ones. By the end of 1925, Hubble had photographed so many galaxies through the 100-inch that he could now classify them according to shape. Hubble's classification scheme took into account virtually every galaxy, from spirals loosely wound to the very tight ones; spirals where a bar crosses the galaxy, separating its center from the curl of its arms; and finally the elliptical systems.

In 1926, Hubble embarked upon his grandest work, the establishment of the redshift as the distance yardstick of the universe. To calibrate the new measurements to those of the Cepheids, Hubble studied the photographic plates in an attempt to find Cepheids in galaxies as far away as he could, while his associate, Milton Humason, took a two-night-long photograph of the spectrum of a distant galaxy.

The result was worth the wait in the frigid dome. When Humason finally studied the plate he saw that the "H and K" lines, the result of calcium in the galaxy, were shifted far to the right. Humason continued to photograph galaxies. Using these photographs, Hubble was now confident that he had calibrated the redshift scale with the Cepheid yardstick. Humason was an extraordinary observer who, years before, had wanted so badly to be part of astronomy that he agreed to be a mule-driver up the steep but beautiful Mt. Wilson road. Years later he would discover a comet that had a most unusual tail.

Redshift is an indication of how rapidly a distant galaxy is moving away from Earth. By May 1929, Hubble understood that the faster an object is receding from us, the farther it is from us, and he was ready to publish his first paper on the relation between distance and radial velocity.

HUBBLE AND SHAPLEY

After Shapley's excellent work with the size of the galaxy, his own ambitions were being thwarted at every turn by Edwin Hubble. "They disliked each other very thoroughly," Shapley's colleague Bart Bok remembered. Shapley even made fun of the English accent Hubble developed when he was a student at Oxford. "I was not like Edwin Hubble," Shapley confided to Bok, "who went to Oxford to put a hot potato in his mouth, and developed an accent which was impeccable."[8]

Personality aside, the conflict between the two scientists reached its peak in 1936 with the publication of Hubble's *The Realm of the Nebulae* in which he summarized his ideas in a book the public could understand. Bok remembers Shapley's foul mood when he got the book late one afternoon, and that his humor was even worse the next morning after he had spent the night reading it. Both Shapley and Bok disagreed with one of Hubble's main lines of evidence for his cosmological principle, that the galaxies were evenly distributed in space. Two years earlier Bok noted from his own observations that the nebulae were distributed unevenly.[9]

As *Realm of the Nebulae* was appearing in bookstores and libraries across the land, another sky survey was taking place by Clyde Tombaugh, discoverer of Pluto. Tombaugh's search did not stop with the discovery of the planet in 1930, and by 1936 Tombaugh had discovered a huge supercluster of galaxies in Perseus and Andromeda. He had also noticed other examples of uneven clustering of galaxies, and voids where there seemed to be few or no galaxies. After reading *Realm of the Nebulae*, Tombaugh felt, using his own observations as evidence, that Hubble was wrong in his idea of an even distribution of clusters of galaxies in the universe.

Using the mighty 100-inch on Mt. Wilson, Hubble had based his data on small samplings of selected areas of sky; Tombaugh's wide-field telescopic camera recorded the whole sky. In the early 1940s Tombaugh met with Hubble, who listened to Tombaugh's argument without much enthusiasm. He tried to argue the point with Hubble, who did not seem interested in discussing it. Tombaugh believed that Hubble did not have much respect for his amateur astronomer background. He did not learn for another half-century that Hubble had had the same argument with both Shapley and Bok.

If Shapley disliked Hubble, the feeling was certainly mutual. Just as *The Realm of the Nebulae* caused Shapley a sleepless night, Hubble defiantly tossed a copy of the *Shapley-Ames Catalogue,* Shapley's careful survey of galaxies, into a wastebasket. In this disturbingly personal period, on August 4, 1937, Hubble discovered a comet near the star Alpha Piscis Austrini.[10]

OF BEGINNINGS AND ENDINGS

The debate about the distribution of galaxies would have to be put off, however. As Hubble and Humason continued to amass data on the expansion of the universe, George Ellery Hale was planning a giant new telescope that could carry Hubble's work even further. Through a 200-inch telescope, Cepheids and redshifts could be calibrated more accurately, and the distance scale could be extended farther out into space. But as the great telescope was nearing completion, war was declared in Europe. With the entry of the United States into the war in 1941, astronomy was essentially shut down as most of the staff astronomers, including Hubble, left for war-related work. With the defeat of the Germans and Japanese in 1945, the astronomers returned to California, where work finally resumed on the dome atop Palomar.

Meanwhile the forty-eight-inch, wide-field camera was completed and working on its maiden assignment, the Palomar Observatory Sky Survey. The seventy-five pages of *Shapley-Ames*, covering 1,249 galaxies, laid a foundation for the Palomar Sky Survey. That forty-eight-inch telescope was capable of spotting galaxies far fainter than could the small cameras used in *Shapley-Ames*, and it solved the riddle about the distribution of galaxies. In his 1954 doctoral thesis, George Abell used the magnificent 14-by-17-inch photographic plates of the just completed survey to record an uneven distribution of clusters of galaxies. On the fine level of detail revealed by the new plates, however, the superclusters, of which the clusters were a part, seemed to be evenly distributed. Hubble's law works—but only on the level of the superclusters of galaxies.

In June 1948, the 200-inch made its celestial debut, and Hubble promptly used it to photograph his favorite close-by variable nebula, NGC 2261. Quickly moving on to more distant targets, Hubble found that he could reach twice as far into space with the new telescope—to a distance he thought was a billion light years away.

In July 1949, while vacationing on a high-altitude ranch, Hubble gave a talk about the universe. Later that night he suffered a major heart attack from which he was lucky to have survived. Four days later, at a hospital in Grand Junction, Colorado, a second attack overtook him, more serious than the first, putting Hubble in agonizing pain. As Hubble recovered over the next months, his main worry seemed to be when he could get back to the 200-inch. At last, in October 1950, Hubble was medically cleared to work on the mile-high Palomar mountain and its 200-inch telescope.

Hubble did get to use the 200-inch a number of times; in fact he took 176 plates with it. On September 1, 1953, the Hubbles were driven to the 200-inch for a three-night observing run.[11] After it ended, Hubble prepared for his next run, which was scheduled for the dark of the Moon period in early October. Meanwhile, on September 28 he talked with Humason about future plans for dealing with the redshift, then left and began walking home for lunch. Grace happened to pass him walking confidently toward home, his eyes ahead, his head in the stars. She stopped and he got into the car. As they continued home, Hubble grew pale, then slumped back into unconsciousness. His last words to Grace, "Don't stop, drive in,"[12] seemed almost as much directed to her as it was his final sailing order for the 200-inch telescope: Keep its motor drive following the distant galaxies, and never stop understanding the cosmos.

THE COSMOLOGICAL PRINCIPLE

If those were indeed his last orders, astronomers have continued to carry them out. Hubble's cosmological principle has been one of the driving forces for astronomical research since that time. If the universe is indeed expanding from a single point in space and time, it should look about the same as viewed from anywhere. This simple Hubble idea is known as the *cosmological principle*. As we understand it now, because of the work done by Shapley, Bok, Tombaugh, and Abell, the cosmological principle works only on extremely large scales of several hundred million light years or larger. Nearby galaxies are not distributed evenly, nor are the clusters of galaxies that lie near our Milky Way. Hubble believed that the clusters of galaxies were distributed evenly throughout the universe. We now know that this even distribution is only at the higher level of the superclusters, or the clusters of clusters, of galaxies. Recent discoveries of a "Great Wall" and a "Great Attractor" containing huge numbers of galaxies indicate, however, that the distribution may not be entirely uniform even at that level.

DOES THE REDSHIFT ALWAYS WORK?

If we knew the value of the Hubble redshift-distance parameter, we would know the age of the universe. Even with the redshift, a supposedly simple yardstick, Nature seems to enjoy throwing curve balls at scientists who think they understand how it works. One such curveball is a beautiful system of five galaxies called Stephan's Quintet. The five galaxies are not particularly faint; I have seen them through my sixteen-inch telescope. In long exposure images, they appear to be joined by a series of filaments. When astronomers have taken routine spectra of each of these galaxies, they found something amazing: the redshift of one of the galaxies indicates that it is only an eighth of the distance of the others. Could that one galaxy really be so coincidentally placed so that it appears to be connected to the others, even though it is really much closer? Or is there a different explanation of its redshift?

An even more startling example of "discordant" redshifts is VV172, a chain of five galaxies. Like Stephan's Quintet, one of the galaxies has an anomalous redshift. If we interpret it in the traditional Hubble way, the fifth galaxy is more than twice as far away as the rest, and by an incredible coincidence it just happens to fill a gap in the chain of galaxies. A third example is the quasar Markarian 205. Although it appears to be very close to the galaxy NGC 4319, in the constellation of Draco not far from the North Celestial Pole, its redshift indicates that it is twelve times farther away! The quasar and the galaxy could be just a chance alignment. But if the galaxy is physically connected with the quasar, and if all the galaxies of VV172 are related, then the anomalous redshift of the quasar must be due to some unknown process other than recession. The answer to these riddles still eludes us.

Besides the Cepheids and the redshift, astronomers have come up with other distance yardsticks. For more distant galaxies, a method called the Tully-Fisher relation works in the same way as the period-luminosity relation of the Cepheids. Instead of relating period of light variation to its luminosity, Tully-Fisher relates the rate of a galaxy's rotation to its total luminosity.

For galaxies at the limit of vision, where they are just too faint to reveal spectra, astronomers have begun to take advantage of supernovae, or exploding stars, to calculate their distances. Take a group of hundreds of tiny faint fuzzy dots, visible through the Hubble Space Telescope. If a supernova explodes in any one of them, it should brighten so that it is as bright as the combined light of all the stars in its galaxy. Supernovae are easy to detect, even in the farthest galaxies, and with each explosion comes an expanding shell of radiation that can also be measured. Observers can compare the expanding shell's growth rate to its speed of expansion toward Earth, and use the result as a distance yardstick.

On a starred night Prince Lucifer uprose.
Tired of his dark dominion swung the fiend…
He reached a middle height, and at the stars,
Which are the brain of heaven, he looked, and sank.
Around the ancient track marched, rank on rank,
The army of unalterable law.

George Meredith, *Lucifer in Starlight*

Two men look out through the same bars:
One sees the mud, and one the stars.

Frederick Langbridge,
A Cluster of Quiet Thoughts, 1896

The Pillars of Creation, showing a new star, with solar system, in formation. NASA/Hubble Space Telescope photograph.

Follow wise Orion
Till you waste your Eye—
Dazzlingly decamping
He is just as high—

Emily Dickinson
c. 1882, Poem 1538

You know Orion always comes up sideways,
Throwing a leg up over our fence of mountains,
And rising on his hands, he looks in on me
Busy outdoors by lantern-light with something
I should have done by daylight,...

Robert Frost, *The Star-Splitter*

The Great Nebula in Orion.
Photograph by Rik Hill of
Tucson, Arizona.

Read Nature; Nature is a friend to truth...
Hast thou ne'er seen the comet's flaming flight?
The' illustrious stranger, passing, terror sheds
On gazing nations, from his fiery train
Of length enormous; takes his ample round
Through depths of ether; coasts unnumber'd worlds
Of more than solar glory; doubles wide
Heaven's mighty cape; and then revists earth,
From the long travel of a thousand years.
Thus, at the destined period, shall return...

Edward Young, *Night Thoughts*, 1741

Comet Hyakutake and a saguaro cactus plant.
Photo by David Levy.

When the Astronomer stops seeking
For his Pleiad's Face—
When the lone British Lady
Forsakes the Arctic Race

When to his Covenant Needle
The Sailor doubting turns—
It will be amply early
To ask what treason means.

Emily Dickinson, c. 1864, Poem 851

This Hubble Space Telescope Deep Space Image shows galaxies farther away than anything shown before. NASA image.

Move eastward, happy earth, and leave
  Yon orange sunset waning slow;
From fringes of the faded eve,
  O happy planet, eastward go,
Till over thy dark shoulder go
  Thy silver sister-world, and rise
To glass herself in dewy eyes
  That watch me from the glen below.
Ah, bear me with thee, smoothly borne,
  Dip forward under starry light,
And move me to my marriage-morn,
  And round again to happy night.

Alfred, Lord Tennyson
Move Eastward, Happy Earth, 1842

I am the owner of the sphere,
Of the seven stars and the solar year,
Of Caesar's hand, and Plato's brain,
Of Lord Christ's heart, and
  Shakespeare's strain.

Ralph Waldo Emerson
The Informing Spirit

Till the sun grows cold,
And the stars are old,
And the leaves of the Judgement Book unfold!

Bayard Taylor, *Bedouin Song. Refrain* 1825-1878

Sky from Jarnac Pond. David Levy photo.

The very source and fount of day
Is dash'd with wandering isles of night.

Tennyson, *In Memoriam,* 1850

Giant magnetic storms called sunspots march across the face of the Sun in this series of photographs by Leo Enright of Sharbot Lake, near Kingston, Ontario.

*As lines, so loves oblique, may well
Themselves in every angle greet:
But ours, so truly parallel,
Though infinite, can never meet.*

*Therefore the love which us doth bind,
But Fate so enviously debars,
Is the conjunction of the mind,
And opposition of the stars.*

**The Aurora Borealis.
Photo by Leo Enright.**

Andrew Marvell, *Definition of Love*

**Does this ageless sonnet evoke the vastness of this
great and distant spiral galaxy? NGC 891 is an
example of a galaxy viewed edge-on from Earth.
Tim Hunter photo.**

*O deep of Heaven, 't is thou alone art boundless,
    'T is thou alone our balance shall not weigh,
'T is thou alone our fathom-line finds soundless,--
    Whose infinite our finite must obey!
Through thy blue realms and down thy starry reaches
    Thought voyages forth beyond the furthest fire,
And, homing from no sighted shoreline, teaches
    Thee measureless as in the soul's desire.
O deep of Heaven, no beam of Pleiad ranging
    Eternity may bridge thy gulf of spheres!
The ceaseless hum that fills thy sleep unchanging
    Is rain of the innumerable years.
Our worlds, our suns, our ages, these but stream
Through thine abiding like a dateless dream.*

Sir Charles G.D. Roberts, *The Night Sky*, 1889

**Above: Scorpius-Sagittarius Milky Way.
Roy Bishop photo.**

The Moon, photographed by David Levy.

*I wandered lonely as a cloud
That floats on high o'er vales and hills,
When all at once I saw a crowd,
A host, of golden daffodils;
Beside the lake, beneath the trees,
Fluttering and dancing in the breeze.*

*Continuous as the stars that shine
And twinkle on the milky way,
They stretched in never-ending line
Along the margin of a bay:
Ten thousand saw I at a glance,
Tossing their heads in sprightly dance.*

William Wordsworth, *Daffodils*, 1807

I will lift up mine eyes unto the mountains:
From whence cometh my help…

Psalm 121

**Twilight at Palomar Mountain Observatory. This view
shows the dome of the two hundred-inch telescope taken
through the open shutters of the 18-inch telescope dome.
David Levy photo.**

HOW OLD IS THE UNIVERSE?

Of the many aspects of Hubble's theory of how much the universe has expanded since the Big Bang, one of the most useful is the Hubble Parameter. Written mathematically as $H0$, it is a measure of how rapidly space is expanding and of how old the universe is. We determine $H0$ with one of the distance scales, of which Henrietta Leavitt's Cepheid yardstick is the most accurate.

Recently, a group led by cosmologist Wendy Friedman used the Hubble Space Telescope to examine the behavior of some Cepheid variables in nearby spiral galaxies. The Cepheid variable evidence tells us that the universe might be younger than 10 billion years old. There is a problem with this relatively young age. Certain white dwarf stars, common in our galaxy, might be as old as 14 billion years in age. Globular clusters might be even older—as many as 18 billion years of age. So how can the universe be younger than some of its stars? Is the universe 10 or 15 billion years old, or somewhere in between? The problem of the age of the universe was one of the prime reasons that the Hubble Space Telescope was built, and in years to come it, or its successors, might provide the answer.

FOURTEEN

1930: CLYDE TOMBAUGH AND PLUTO

In my stars I am above thee; be not afraid of greatness. Some are born great, some achieve greatness, and some have greatness thrust upon 'em.
—Shakespeare, *Twelfth Night*[1]

Dr. Slipher, I have found your Planet X.
—Clyde Tombaugh, February 18, 1930

February 17, 2001, was a special day for the city of Las Cruces, New Mexico. On that day the local Unitarian Church dedicated a spectacular stained glass window that portrays the life and accomplishments of Clyde Tombaugh, who discovered the planet Pluto in 1930.

The window is a treasure, an inspiring look into the heavens that might be unique in the annals of astronomy. For the 200 colleagues, friends, and family gathered there, it offered a chance to reflect on the life of this particular cosmic discoverer. What no one knew that afternoon was that far out in space, a distant star that Tombaugh also discovered wanted to participate as well. It was TV Corvi, or Tombaugh's star in Corvus. (See chapter 2.) The following evening, I peered through my eight-inch telescope and saw the star undergoing a major outburst; in just a few hours it had risen at least 100 times in brightness. That night of February 18, the night Tombaugh's star exploded in celestial fireworks, was also the seventy-first anniversary of the discovery of Pluto.

FOR SPACIOUS SKIES

After sailing out beyond the planets, past the stars of our own galaxy, to the very edge of the universe, it might seem a letdown to have to return home to our own solar system to continue our story. But as one group of astronomers pushed their knowledge to the very edge of the universe, another continued the effort to understand our own neighborhood. Both stories are important and full of discovery, and as we shall later see, in the turn-of-the-millennium work on planets around other suns, the two forces begin to merge.

Vesto M. Slipher, as we have seen, left to Edwin Hubble his groundbreaking work into the redshift of the nebulae. Slipher's boss, Percival Lowell, was interested more in the planets, and especially in the possibility of life on Mars. Although he always believed that Martians had overcome regional differences to unite and construct a system of canals to bring scarce water to sparse areas of their world, he was painfully aware that much of the rest of the astronomical community was ridiculing him and his observatory on the Flagstaff mesa called Mars Hill. He therefore thought of a new area of research, dating back to the discoveries of Uranus and Neptune. Like that of Uranus, Neptune's orbit also seemed strange. Lowell concluded that another world, a "Planet X" lay beyond the distant blue world of Neptune.

In 1905, the year he began his search for Planet X, Lowell actually photographed the world he sought but never knew it. After he died from a stroke in 1916, his observatory was thrown into disarray for more than a decade while his widow, Constance, fought for control of his estate. Late in the 1920s, the legal issue was finally settled and Lowell Observatory director V. M. Slipher could get back to the business of doing science at the observatory. The trustee, Roger Lowell Putnam, firmly believed that his Uncle Percival would have wanted the search for Planet X to continue, and so he asked Slipher to arrange for the construction of a new and powerful patrol telescope. Realizing that the tedious work of taking the photographs for such a search was not for him, Slipher wondered about hiring someone to do the job in Flagstaff's frigid nights.

It was just about that time that Clyde Tombaugh decided to give up farming. It would no longer be his career.

A FARMER AND AMATEUR ASTRONOMER

Born February 4, 1906, Clyde William Tombaugh grew up on farms in Streator, Illinois, and Burdett, Kansas. While operating the farm machinery of the time, he occupied his mind with mathematical puzzles beyond the reach of most of us. How many cubic inches were in Betelgeuse, the bright red star in Orion, was one such problem. The answer, in case you too have been wondering, is 1 duodecillion, or 1 followed by 39 zeros.

Tombaugh's interest in astronomy stemmed from elementary school, where he wondered what the weather might be like on other planets, particularly on Mars. The red planet had been in the news off and on for the preceding quarter century, when Percival Lowell suggested that it might be inhabited. A few months after Clyde graduated from high school in 1925, a lead article in *Scientific American* caught his attention. "The Heavens Declare the Glory of God" was an article about what amateur astronomers were doing to bring down the cost of telescopes. In the town of Springfield, Vermont, amateur astronomers were grinding their own mirrors, completing the construction of their own telescopes, and even having an annual conference called Stellafane, or Shrine to the Stars.[2] (Stellafane still takes place each summer at the same site, traditionally at the first dark of the Moon after the Green Mountain corn is ripe.)

The article in *Scientific American* sparked a revolution in astronomy, as many amateur astronomers around the world started to make their own reflector telescopes. Tombaugh was not happy with his first efforts, but in August 1927 he purchased the optics for a nine-inch reflector which turned out beautifully and which he used for the rest of his life. "I struggled with the figuring of that mirror for many weeks, and by trying to get one [error] zone out I'd generate another. Finally I got them all licked. That really paid off, for it would take magnifying powers as high as 400."[3] For the mounting, Tombaugh used parts from his father's old 1910 Buick and a cream separator.

By 1928 Tombaugh was well into using this new telescope. June 10 was a hot, muggy day. By mid-afternoon, storm clouds were gathering in the northwest. The family hurried indoors as torrents of rain pelted the farm. Then for fifteen terrible minutes, dense hail destroyed the family's entire crop of wheat and oats. The family income was completely wiped out, and that day the young Tombaugh decided he had had it with farming. That was the day Tombaugh gave up on agriculture as a career.

FROM THE FARM TO THE OBSERVATORY

Uncertain if he could make a living in astronomy, in the autumn of 1928 Tombaugh wrote the proprietor of a telescope company in Wichita, Kansas, and he also wrote to V. M. Slipher, the director of the Lowell Observatory. Slipher's reply, on November 30, was more encouraging than the young Tombaugh had expected:

> Your letter [the director wrote] was duly received and we have read it with a great deal of interest. You appear to have had a good deal of valuable experience in the

construction of reflecting telescopes and in securing satisfactory performance with them. In this connection we would appreciate your sending us some drawings you have made in your observations of Jupiter and the other planets. We note that you are interested in securing employment at an observatory. We shall have need a little later of some help. . . .

Are you in good physical health? Are you single or married? We understand from your letter that you are intending to come west late in December. May we hear from you, however, by letter at your early convenience.

We note that you feel particular interest in the study of the planets. You may—or may not—know that this Observatory is particularly concerned in the study of planets.[4]

Tombaugh reacted excitedly to this turn of events. Could it be that Lowell Observatory might in fact rescue him from the farm, and offer him a rare coveted position in astronomy? He quickly assembled some drawings of Jupiter and Saturn, on which were clearly noted their time and date so that the observatory staff, particularly the director's brother Earl Slipher, could compare them with their own photographs of the planets taken at the same time.

Tombaugh received a most encouraging Christmas present from Lowell Observatory in the form of a letter, dated December 21, from Director Slipher: "Your detailed letter of December 3, and the planetary drawings have been examined with interest. Evidently you have succeeded very well, both with the telescope and with observations with it." The letter continued with a specific hint of what Tombaugh's future might behold:

We expect to have soon a new photographic telescope for some special work and shall need someone to operate it. . . . It is perhaps possible that you might be able after some instruction here to be able to make exposures with this instrument so we are thinking of you as possibly developing into an assistant for that work. It would mean, as you can well imagine, long nights at the telescope during the moonless nights. Still this is something that one who undertakes astronomical observations must expect.[5]

With this letter, Tombaugh was obviously optimistic about the possibility of work. He wrote back to assure Slipher that he was interested in the photographic task and capable of learning how to do it. On January 2, 1929, Slipher responded with the offer that would, within fourteen months, lead to the discovery of the solar system's ninth planet:

Your letter of December 28 has been duly received. And we note your eagerness to get into actual observatory work. We are more in need of help now than ever, and we are willing to have you come here on trial, for a few months at least. . . .

"I am sure I do not need to tell you that there is lots of hard work and a good many long and some uncomfortable hours at night with the instruments. But if one is interested in the work and is not averse to working hard there are compensations in it aside from the living it affords. There are other things to be sure which one can do and find better pay; but to us interested in astronomy they do not have the satisfying interest that we find in the study of this science.

You may come as soon as you conveniently can. If you need money advanced to meet traveling expenses you telegraph us if desirable to save time. . . . I thought you might care to bring—if convenient—some bedding and linen with you." [6]

FIRST SEASON AT LOWELL

Tombaugh did not take Dr. Slipher up on his offer of a free ticket. He invested in his own one-way ticket from funds earned while running his neighbor's combine. On January 15, 1929, just thirteen days after Slipher's letter, Tombaugh was met by the director at the Flagstaff rail depot. The telescope was not quite ready yet, but until it was Tombaugh learned the ropes of the observatory. He also learned what his immediate future would entail: He was to take glass photographic plates with the telescope as part of a search for "Planet X." Calculated by Percival Lowell years before, Planet X was the body that he believed to be affecting the orbital motions of Neptune and Uranus. A month later, the thirteen-inch lens arrived. With some uneasiness, Slipher, Carl Lampland, and Tombaugh took a thirty-minute exposure of Orion's belt and sword region. The result was sharp images right across the plate—the telescope, it seemed, was ready to begin to search the sky.

Tombaugh spent the next few weeks working out some minor problems with the telescope, and using the new blink comparator with which pairs of identical plates, separated by a few days in time, would be scanned. On April 6, 1929, all was ready for the search to begin.

Actually the search was not beginning, but resuming after a long break. July 2, 1916, was the last exposure in the previous search with a different telescope. That last log entry on that date was "Lunch," in the hope that the break in photography would be resumed later.[7] That break lasted thirteen years, and when the search resumed, there were problems. The early plates Tombaugh took contained bad news—planet suspects appeared everywhere, and it was difficult to know, by the amount of motion from one photographic plate to another, just how distant the unknown point of light was. From April 6 to the time work stopped for the summer storm season in June, Tombaugh took about 100 one-hour exposures. Both V. M. and E. C. Slipher blinked some of these plates for almost a week in hopes of making a quick find of Planet X. The brothers were quite depressed that the new telescope had yielded nothing.

On June 18, the director spoke to Tombaugh. From this point on, he said, the blinking of all plates, as well as the actual photography, would be Tombaugh's responsibility. Tombaugh was stunned at this turn of events and wasn't sure how to interpret it. Either the older men now had confidence in Tombaugh's ability and discipline to do the blinking of his own plates, or they had concluded from their earlier blinking that Planet X was unlikely to be found and that the search should be given lower priority. The answer was probably the former reason. Lowell's trustee, Roger Lowell Putnam, had put his own influence and reputation on the line to raise the funds for this telescope, and he often wrote the Sliphers about "Uncle Percy's wishes and to carry on his work."[8] Planet X was high on Putnam's list, and Tombaugh knew that the success, or failure, of the project would now be up to him and the sky. The weight of so much responsibility, coupled with some homesickness, depressed him, and Carl Lampland wisely suggested that the director permit Tombaugh to return home for a summer break.

It was a good idea. Back home in Kansas, Tombaugh had a chance to rethink the entire survey. His biggest concern was even if he did find the planet, how would he know? With some 300,000 stars in each of the Milky Way-rich plates in Gemini, where the planet was believed to be, any faint moving object could be an asteroid, or it could be Lowell's Planet X. While busy at the farm that summer, Tombaugh went back to his old practice of making mental calculations while working farm equipment. Only this time, he wasn't figuring out the number of cubic inches in Betelgeuse. During the summer of 1929 he thought about the observing problem he faced. There had to be a way to separate the appearance of relatively close-by asteroids with that of a distant planet.

The major planets orbit the Sun, he thought, at different distances. As Mars, the asteroids, Jupiter, Saturn, Uranus, and Neptune all orbit the Sun, they appear to move eastward through the zodiac, crossing the sky once each orbit. But as the Earth overtakes these outer worlds, they appear to slow down and reverse direction for a few months. It is the same effect seen by a driver passing a slower automobile; the other vehicle appears to move backwards at the moment of overtaking. This retrograde motion takes place when the planets are near opposition to the Sun.

The solution seemed so simple, yet so elegant. Why not take all the search plates at opposition? With that strategy, any solar system object will be retrograding at a speed in reverse proportion to its distance from the Sun; the farther it is, the less the shift across the background of stars. Most of the asteroids moved at 7 millimeters per day at the scale of these plates. Jupiter moved more slowly, Saturn slower yet. All he had to do was to find an object that moved a half millimeter per day, and that would be the trans-Neptunian planet.

SECOND SEASON

When Tombaugh returned that fall he was reenergized, bristling with confidence. If the planet were there, and if it were brighter than the seventeenth-magnitude limit of the telescope, Tombaugh should find it. It was not until many decades later that Tombaugh learned that Lowell himself had also come up with the opposition strategy, but that somehow the Sliphers and Lampland had forgotten about it, and they enthusiastically endorsed Tombaugh's plan. In any event, in the fall of 1929, with clearing skies after the summer clouds and rain, Tombaugh began taking his series of hour-long exposures, moving methodically through the sky, hovering in the opposition region. He repeated each photograph days later, and again a few days after that. On January 21, 1930, he began an hour exposure guided on the star Delta Geminorum. After a few minutes a biting wind sprung up from the northeast. As Delta swelled almost beyond recognition in Tombaugh's guidescope, he struggled to complete the hour-long exposure. Two nights later, he repeated the picture under much better conditions, and on January 29 he took a third plate.

Even though each exposure took an hour, and time was needed to unload the plate and load a new one and take the telescope to a new field, Tombaugh was taking plates faster than he could blink them. Each pair of plates was placed into the blink comparator and once the stars were perfectly aligned, Tombaugh would turn on the device that automatically allowed him to inspect a tiny area of one plate, then the same tiny area of the other plate. Slowly he would cross the plates until, days or weeks later, he finally finished blinking the pair.

THIRD SEASON

In early February the area of sky he was photographing was in the winter Milky Way in Taurus, an area so rich with stars that Tombaugh would spend four or five *weeks* blinking a pair of plates. To cover more area faster, Tombaugh thought to skip over the Taurus plates for the time being, and tackle the less demanding plates centered on Delta Geminorum that would take a small fraction of that time. Since Gemini was the region of sky that Lowell had most favored for his planet, this plan met with the director's approval, and on February 18 Tombaugh placed the second and third Delta Geminorum plates into the comparator.

Throughout that morning, Tombaugh's blink comparator went *click, click, click* as its electromagnet shifted the optics from one plate to the other, and it was loud enough that Carl Lampland, from his office across the hall, knew that Tombaugh was busy patrolling the sky. He had to take several breaks throughout the day as his eyes tired from the strain. At one o'clock Tombaugh returned from lunch to start again. Throughout the afternoon he continued blinking until about

3:50, when he encountered brilliant Delta Geminorum, his guide star. It was now four o'clock, and outdoors the sky was already darkening. Delta Gem moved away as Tombaugh moved a field away, then another, and yet another. Through the monotonous search through distant stars, he almost missed a faint starlike object appear, then disappear, on one of the plates.

"I saw a little image popping in and out, looked to one side, and saw another one [on the other plate] doing the same thing." They seemed to be out of phase with one another, so that when one appeared the other disappeared. Turning off the automatic blinker, he wondered if they were defects in the plate—just artificial spots in just the right places to fool a careless observer. Almost a week of time had separated the two images, enough to separate them considerably. Across the hall, Lampland noted that his young colleague had stopped blinking.

Tombaugh measured the shift in position, and it was perfect for a planet way beyond Neptune. *If only the images were real . . .* He inspected the plates manually to make sure the object was moving in the right direction. "I took note of the dates of the plates. The image on the earlier plate was to the east, the one on the later was to the west; it was retrograding!" Next came a check to see if these could be two independent variable stars. Tombaugh replaced the plate from the 23rd with the poor quality one taken in the wind of January 21. If that plate showed the image shifted further to the east, Tombaugh thought, "that would pretty well clinch it." The plate was not nearly as good as either of the two later ones. With trembling hands Tombaugh aligned the plate, then looked through the eyepiece. The object was there, right where it belonged. "I was walking on the ceiling," Tombaugh recalled. "I was now 100% sure."[9]

Pacing the room with excitement, Tombaugh remembered one more possible check he could perform. As a backup on the larger telescope's pictures, Slipher had mounted a secondary camera whose small plates had been exposed as well on the Delta Geminorum region. Could those plates also show the new object? It took him some time to locate the images on these plates, with their different scales. But the images were there. "I can hardly describe to you how intense was the thrill I felt. I was looking through my hand magnifier identifying the configuration of the stars. I could hardly see them. My hand was shaking; I was literally shaking with excitement."[10]

It was just past 4:45 that afternoon. Tombaugh crossed the hall to Lampland's office and blurted, "I think we've found Planet X." Lampland had been sitting in his office for the past forty-five minutes. "I heard the clicking sound and then this long silence," Lampland told him later. I had suspected you had found something."

While Lampland was examining the plates, Tombaugh left the room again. He walked down the long hallway with its tall ceiling, past the secretary's office, and then he stopped just outside the office of the director. He straightened up, trying to calm himself as he walked through the open door. Slipher was reading some papers, and looked up as the young man entered.

Tombaugh took a deep breath. "Dr. Slipher—I have found your Planet X." Slipher rose from his chair and stared at his young assistant. "I'll show you the evidence."

Slipher darted down the hall and burst in on Lampland, who was studying the images. "It looks pretty good," he said to Slipher. Slipher examined the images and agreed. The three men then talked for the next hour about the implications of the discovery. The sky was now completely dark. Slipher put his two colleagues under a gag order—no one was to know. Tombaugh took Slipher's admonition so seriously he didn't even write his parents.

That evening Tombaugh descended Mars Hill. With no sign of clearing he decided to take in Gary Cooper in *The Virginian* at the local theater. Once back at the observatory, Tombaugh went into Lowell's library, looking at the old books, and thinking about the long-deceased founder's hopes for his observatory.

The following night the sky cleared enough so that Tombaugh could rephotograph the field. From this new plate he made a small, 5-by-7-inch contact print showing the area that included both Delta Geminorum and the new planet. On February 20, armed with this print, Slipher, Lampland, and Tombaugh walked to the old twenty-four-inch telescope, the one that Tombaugh, years earlier, had dreamed of looking through someday. On this night they did indeed look through the big refractor, at a new world.

A NEW WORLD IS ANNOUNCED

By early March, the observers, now joined by E. C. Slipher, had noted that the planet's motion still conformed to that of a world beyond Neptune. Slipher was in regular contact with Harlow Shapley, director of Harvard Observatory, whose job it was to publish the announcement. They selected March 13, the anniversary of William Herschel's discovery of Uranus, as the date.

The new planet made front pages all over the world. Instantly the observatory atop Mars Hill was famous around the world, as was the young farmer-turned-observer who had discovered it. "This is one of those discoveries," Lowell trustee Roger Lowell Putnam wrote Tombaugh on official Estate of Percival Lowell stationery, "that has not happened by chance, but has been the result of painstaking, hard work, much of it monotonous. I know how pleased you must be and having this work bear fruit so quickly, and I think you know without my telling, how pleased I am. I am looking forward very much to seeing you when I come out in May."[11]

"As a Kansan," wrote the chancellor of the University of Kansas, where Tombaugh would later register as a freshman student, "I wish you to know of our pride and satisfaction in the significant contribution you made to the discovery of the new planet. I hope at no distant time that you may find it possible to visit the University and allow us to express to you in person our pride in your achievement."[12]

But these were official letters. One can imagine the deep feeling of pride that his parents felt. "Dear son, his mother wrote, "Well! well! well! Just 14 mo. ago last Fri, you left for Flagstaff, and about 9 o'clock on said Fri. morning we heard the *grand good news*. . . . Telephone calls came so thick for awhile that they almost swamped us and we rushed around to different photo albums and boxes of pictures to find what was wanted. . . . Goodness it took us off our feet!"[13]

Clyde's father was no less thrilled. "Dear Clyde," he wrote,

Well, well, you old headliner, how does it feel to be a hero in the public eye? Ha, you sure took us by surprise keeping mum about your find until it was proven a fact.

Friday morning we were sitting around the house after breakfast when the phone rang, Tiller and Toiler [their local newspaper] calling, "Did you know, your son has discovered a planet?" [The letter continued with the details of reporters who called for interviews and photographs, and then] Planted potatoes Sat. AM. and all went . . . for a little celebration in your honor. . . .

Well you must write and tell us all about it, even if you have to put on a few extra stamps. The old town feels that you have put them on the map, and although the discovery of new planets is something over the heads of most people in country towns, they are proud of you in a quiet way, and glad that you have made good.[14]

BEYOND PLUTO

In his own letter to the *Daily Science News Bulletin*, Tombaugh wrote prophetically that "the work on the planet, however, is far from finished. Now that it is found, the elements of its orbit, and much else concerning it, must be learned, so doubtless it will be a much observed object."[15]

Much observed indeed, right up to the present. In late spring 1930 the new world was named Pluto. (Walt Disney's floppy-eared dog, curiously, acquired the same name just a few months later.) In 1978 James Christy, observing from Flagstaff's U.S. Naval Observatory, discovered Pluto's moon Charon. Even today the planet is bathed in controversy and interest. It clearly was not the Neptune-sized world that Lowell had thought to be affecting the motions of Uranus and Neptune, for its mass is too small. In fact, no unknown world is affecting those two planets. Now that we know their positions more precisely, no new world need be invoked to explain their orbits.

As our understanding of the outer solar system increased, we now see Pluto as a very different object. It has a moon, and orbits the Sun in the midst of many tinier comet-like worlds called the Kuiper Belt. Meanwhile, Tombaugh attended the University of Kansas where he earned his bachelor and master degrees, and he married Patricia Edson, an artist, in 1934. At Lowell he went on to make other discoveries, including a comet, many asteroids, half a dozen star clusters, and a new variable star in the constellation of Corvus. In 1945, Tombaugh was asked to leave Lowell, because, said Slipher, the observatory did not have the funds to keep him. He moved to Las Cruces where he used his knowledge of telescopes to develop the first missile tracking system at the dawn of the space age, at the White Sands Proving Ground. His career continued at New Mexico State University, where he started the most intensive planetary observation program ever undertaken. He died at his Las Cruces home in 1997.

SOME NOTES ON THE STATUS OF PLUTO, AND ITS DISCOVERER

In recent years, some astronomers have questioned whether Pluto is large enough to even be considered a planet rather than an asteroid. Pluto is every inch one of the nine major planets in the solar system. It is far larger than the largest known asteroid, and when it was formed, it was large enough to condense under its own gravity to form a sphere, instead of an asteroidal body of irregular shape. Pluto is composed mostly of ices, unlike the rocky structure common to asteroids. Incidentally, Pluto has much more in common with the Earth than Jupiter does. Were we to stand on Pluto we would see a night sky like our own, and a single large moon. We couldn't even stand on Jupiter, you would fall right through it!

On February 3, 1998, one day shy of what would have been Clyde Tombaugh's ninety-third birthday, Johannes Anderson, the General Secretary of the International Astronomical Union, announced that Pluto would not be assigned an asteroid number, and that no proposal to change the status of Pluto as the ninth planet in the solar system has been convincingly made. Let the word go out that there are nine planets in our solar system, and that Pluto rightfully belongs as one of them.

Pluto is unique, and its discoverer, who died in 1997 at age ninety-one, also was unique. When I first met him in 1963, I could not believe I was in the presence of one of a handful of people who had discovered a new planet. As I got to know him better, I saw a man committed not only to his science but to the strange word plays of the English language. His puns were legendary: He would start a lecture with "For 50 years I've been a Plutocrat!" and when a balky missile refused to work at White Sands, he suggested they "fire it."

Clyde Tombaugh was one of the last great observers, and never lost his love of observing. A few years before his death, he rebuffed an attempt by the Smithsonian Institution to obtain his telescope for their collection with the remark that he was still observing with it. In professional observatories today astronomers huddle not in the cold but around computers in warm rooms while their tele-scopes gather data in silence. His world, and his time, will be missed.

FIFTEEN

1948:
BART BOK AND
HIS GLOBULES

"When you are in an observatory at three o'clock in the morning, stop your photograph," he would say. "Stop your photometer. Walk away from the telescope. Walk down the stairs. Walk out the front door. Now walk twenty paces, no more, no less. Then stop—and look up at the sky—just to make sure you are making bloody sense."
—Bart J. Bok, advising his students

I will never forget the day Bart Bok told me how he became interested in astronomy. He recounted his story shortly after I met him, then in retirement at his home. Although he was then seventy-three and had a weakened heart, he talked with emotion, humor, and an animation rare among the scientists I knew. He was in his element, sitting on a easy chair and chatting amiably about how astronomy had changed during his lifetime. But Bok's role in that change had to start somewhere. When I asked him how he got started, he talked of how, as a twelve-year-old Boy Scout patrol leader in Holland in 1918, he went camping with a group of other patrol leaders and their supervisor. They were all looking at the stars, and as the supervisor tested the youngsters on their knowledge of the constellations, it turned out that one patrol leader knew much less than the others. " 'Look at Patrol leader Bok,' the supervisor chided him, 'with his patrol at night as they sit around the campfire, and it's a beautiful night out and the Milky Way is out too.' And one of the little boys says, 'Patrol leader Bok, what is that star?' Since our hero Patrol leader Bok cannot answer, he takes a log and puts it into the boy's mouth to quiet him for a while. Then a second little boy asked, 'Patrol leader Bok, what is that star over there?' And once again, Patrol leader Bok has no idea, so he puts a log into the second boy's mouth. After a while, when everyone should be learning about the stars, Patrol leader Bok has eight little boy scouts sitting around the fire with logs in their mouths."[1]

That did it for Bok. "David, I was so humiliated by what the supervisor told me that I resolved immediately to learn the constellations. And that is how I got interested in astronomy."

From that austere beginning Bok built his interest gradually. He bought a book on astronomy by the Nova Scotia-born astronomer Simon Newcomb.

LEIDEN

In 1924 Bok entered the University of Leiden as an undergraduate. The grand jewel of the ancient city of Leiden in western Holland, the university was a lively and competitive place of learning. On one of his first days there, Bok decided to explore the treasures in the physics reading room. Also looking over the physics books that day was another first-year undergraduate named Gerard Kuiper. That two young students would meet this way was actually a rare event: At the time Leiden was in the habit of admitting a new astronomy student every three years; this year it admitted two. The friendship that began between the two young men that day would last a lifetime, but in its first words lay the seeds of the competitive spirit that characterized it over the years. Bok talked excitedly about his interest in the Milky Way, and Kuiper called the Milky Way "an interesting minor field," saying his interests in the solar system were much broader. After the time at Leiden, Bok and Kuiper went their own ways. Kuiper would propose the existence of a band of comets orbiting the solar system beyond Neptune; this Kuiper Belt would become his hallmark contribution. Meanwhile Bok would conduct a lifetime of study of the Milky Way. The two scientists reunited at the University of Arizona in the mid-1960s where Bok became director of Steward Observatory and Kuiper the founder of the Lunar and Planetary Lab across the street. The University of Arizona acted as the final arbiter on their relationship at the end of the century. In 1996 it named their largest telescope on Kitt Peak after Bok, and three years later it honored Kuiper by naming its sixty-one-inch reflector in the Catalina Mountains after him.

These honors, of course, were in the distant future, long after both were dead. At Leiden, Bok became fascinated with Shapley's work. As we learned earlier in chapter 7, Shapley erased our notion that we were at the center of the galaxy; "we are now outcasts," Bok explained, "at the galaxy's edge."[2]

After completing their third-year examinations, Bok and Kuiper, and two friends, decided to travel to Norway to see an eclipse of the Sun. They pedaled by Elsinore, where they saw Hamlet's castle, and by June 29, 1927, they arrived at the bank of the Hallingdal River. Kuiper and Bok had prepared an experiment designed to test the true color of the Sun's corona by shining lights on disks of different colors and then choosing the disk that most closely matched the color they perceived as that of the corona. Lights would have pointed away from the observers so as not to disrupt their viewing of the corona. Unfortunately, the eclipse they had come so far to see was itself eclipsed by heavy clouds. "The shadow of the Moon came very quickly over," Bok recalled decades later, "and we saw the darkness quickly approaching across the snow on the mountains. That was the most glorious sight."[3]

Actually, Bok did not miss this eclipse at all. This was the first total eclipse of the Saros series, no. 145, and the eclipse repeated itself eighteen years, eleven days, and eight hours later on July 9, 1945, when the Moon's shadow hit the Earth in the western United States and climbed through central Canada, Hudson's Bay, Greenland, and northern Europe. The eclipse took place again on July 20, 1963, when it crossed through Alaska and Canada—I saw it from southern Quebec—and again on July 31, 1981, when a much older Bart Bok finally got to see it, this time in Russia. (Both of the authors saw the eclipse when it returned on August 11, 1999.)

THE IAU MEETS IN LONDON

In the fall of 1927, Bok began graduate studies at Groningen University. Bok's research plan was at first a study of the structure of the Milky Way Galaxy. He also enthusiastically joined in the planning for the major event that would come with the summer of 1928—the triennial Congress of the International Astronomical Union. This would be Bok's chance to meet the astronomers of the world, especially Harlow Shapley. The meeting worked out well for Bok. Shapley was impressed with the young graduate student and offered him a fellowship to come to Harvard after he completed another year in Groningen.

Bok met someone else during that meeting. He and Kuiper were assigned to the reception committee, and as the train carrying the delegates arrived, Bok noticed a young woman emerge from it. Her name was Priscilla Fairfield. "I was on the reception committee," Bok said, "and I received Priscilla as she came off the train." He received her so well, in fact, that at the end of the ten-day general assembly he asked her to marry him. At first she put off his proposal, but in the following months he persisted, and finally she accepted.

In the late summer of 1929, Bok left for the United States, where his first priority was to meet Priscilla's family. Wanting to make a good first impression, he worried about his clothes not being the best, but at least he knew his shoes would be nicely shined. Following the custom of European hotels, Bok left his shoes outside his room door so that the following morning they would be returned, shiny and fresh. At sunrise, when Bok opened his door, there were no shoes! A horrified Bok found out that it was this hotel's custom that anything left outside a room was to be tossed out. After a frantic search by the hotel staff, the shoes were found and returned.[4]

BOK, SHAPLEY, AND HARVARD

The young Harlow Shapley had a far different basis in astronomy from that of Bok. Shapley began his career not in academics at all but as a reporter for a newspaper. By the time he entered the University of Missouri, he had become disillusioned with the journalistic lifestyle. Reading through the list of courses in alphabetical order, he rejected archaeology because he couldn't pronounce it. Since he could pronounce the next course on the list, astronomy, he chose the heavens as his career. (We have seen in chapter 7 how Shapley completed his work on the Cepheid variables.) By 1929 he had been director of Harvard College Observatory for almost a decade. Shapley believed that the work of the observatory would be more effectively pursued if the staff had fun together as well. Shortly after Bok arrived at Harvard, Shapley found an old libretto hidden away. It was an astronomical spoof of Gilbert and Sullivan's *H.M.S. Pinafore*, and Shapley enthusiastically proposed that the entire staff produce it for the occasion of the American Astronomical Society's meeting at Harvard. It was during the many rehearsals for the operetta, in which Bok played the role of the Man of Providence, that Bok and Shapley really got to know one another.

The Harvard Observatory Pinafore was performed on December 31, 1929. It was a delightful play, filled with astronomical stories and observing disasters. Bok was already learning the perils of getting a recalcitrant observatory dome to start turning, rolling a large observing chair across the floor without hitting the telescope, and measuring starlight with a device called a photometer . . .

(Tune: When I was a Lad)

I turned the dome with so grand a shock
That I broke two windows and the Elliot clock;
I burst the gas pipe rolling the chair
And created a blaze for the winter's scare.
For my worthy zeal they requested me
To try my strength on Photometer P.[5]

. . . and was already becoming familiar with what it would be like to be an observer:

(Tune: A British Tar)

An astronomer is a sorry soul,
As free as a caged bird;
His sympathetic ear should be always quick to hear
The directorial word.
He must open the dome and turn the wheel,
And watch the stars with untiring zeal,
He must toil at night though cold it be,
And he never should expect a decent salaree.[6]

Shapley was genuinely excited about producing the play for the national meeting of the American Astronomical Society. He intended to allow it to bring the hardworking observatory staff together as a family. For a young graduate student freshly arrived from Holland, there could be no better introduction. Shapley and Bok became very good friends after that. And despite his initial reluctance, Bok went on to become one of the observatory's most enthusiastic supporters of observing and observers. One night, for example, student Arthur Hoag had observed the entire night before finally heading for bed. "The following day," he recounts, "I was gently wakened out of a stupor by the distinguished professor of astronomy. Dr. Bok was shaking me, gently, and saying, 'Your breakfast is ready, sir.' "

A NEW COMET

On a summer night in 1949, Bok was training one of the younger observers, Vainu Bappu, to use the twenty-four-inch Schmidt camera. (Bappu would later become president of the International Astronomical Union.) As Bok and Bappu took one photographic plate after another with the telescope that night, Bok allowed Bappu to gain experience in the art and science of astrophotography. The next afternoon, they took the plates into the darkroom and processed them. Bok recalls what happened when Bappu's first plate emerged from the fixer: "Vainu and I looked at the plate, and the images were perfect. He had focused correctly, done everything right, nothing wrong. It was a beautiful plate."

At this point Gordon Newkirk, another undergraduate student, walked in. "You must have a look," Bok said proudly, "at Vainu's first beautiful plate!" Newkirk studied the plate briefly. Almost casually he asked, "Hey, what is that fuzzy line there? Is that a flaw on the plate?"

Bok studied the plate carefully using a small hand magnifier. There was a fuzzy streak with a protrusion off in one direction. He strongly suspected that the fuzzy line was the streak left by a moving object. "Gordon, that is no funny plate defect. That blob there is a bloody comet!" After confirming the comet on a patrol plate taken earlier, Bok telephoned Shapley and reported the discovery. Once he had confirmed that it was not an already known comet, Shapley announced the find as Comet Bappu-Bok-Newkirk.

Vainu Bappu was thrilled with his discovery, particularly since no comet had ever been found in historical times by a native of India. "I would give anything to find one," he had told a colleague. On the very first photographic plate he ever took, he found his dream.

DARKNESS IN THE MILKY WAY

Although our galaxy contains hundreds of billions of stars, it also carries along an incredible amount of nonstellar material in the form of clouds of hydrogen. As Bok studied the galaxy he suspected that these clouds contained precious clues about how stars are formed. He became interested in these clouds while writing his doctoral thesis on Eta Carinae, a star surrounded by nebulosity. "The grand sweep of the swirling gases" was even more interesting to him than the star itself.[7]

When John Herschel traveled to Africa in the 1830s (see chapter 9), he observed a patch of sky near the Southern Cross that seemed devoid of stars. He saw areas even more devoid of stars than this "Coal Sack" structure. He called them holes in the heavens, and two generations later they attracted the interest of Edward Emerson Barnard. Barnard carefully catalogued these dark regions. Rather than seeing them as clear windows to something beyond the galaxy, Barnard interpreted them as clouds of dust that, unlit by nearby stars, blocked our view of the stars behind them. In his own dissertation, Bok reported the existence of a dark nebula he thought might be close to 3,000 light years away.[8] A few years later, in his first published book, he described the numbers, sizes, and studies of all known dark nebulae in the Milky Way.[9]

Shortly after the end of World War II, a young woman named Edith Reilly was hired by Harvard College Observatory as a technical assistant. One afternoon she walked into Bok's office and asked him if she could assist in some way with his studies of dark nebulae. At the time, Bok was hoping to develop a classification system for different shapes of dark nebulae. Bok knew that since Reilly had multiple sclerosis, she would not be able to lift the 8-by-10-inch plates and set them under a lens for examination, but she did have the enthusiasm to explore and search for the dark nebulae, which would appear as bright spots on the negatives.

Bok had the perfect solution. Reilly could examine Barnard's old catalogs of several hundred dark nebulae. Based on Barnard's careful descriptions, Reilly could divide them into groups. Although the work was routine at first, Reilly soon noticed description after description of small, round, and very dense nebulae. Soon Bok was entranced with these small dark nebulae, and decided to focus his energy on them. Using the Jewitt Schmidt camera at Harvard's Oak Ridge Observatory, he photographed the objects Reilly had picked out of Barnard's catalogs, and with Reilly published a preliminary paper about them in 1947.[10]

The 200 or so dark clouds Bok and Reilly found are typically round, about one-sixth the apparent diameter of the Moon in the sky. They are extremely dense, so thick that if one of them were to cover the zero-magnitude star Vega, even the Hubble Space Telescope, which can penetrate down almost to thirtieth magnitude, would not be able to spot it. "Through a telescope you could come to the leading edge of one of these things and suddenly the stars would just disappear," he told me. "And then you would push the telescope's slow motion button a bit and bloop! The stars come back."[11]

These clouds, Bok believed, marked the birthplaces of new stars. After staying together in space for untold millions of years, they begin to collapse under their own gravity. As they get denser the process of contraction continues until the center ignites, stellar fusion begins, and a new star is born. Bok and Reilly had stumbled onto a gem of stellar physics.

As Bok tried to locate more of these objects and study them further, he also set about looking for a suitable name. The answer came to him one morning as he opened the front door of his home to pick up the freshly delivered glass bottles of milk. In those days, milk was not homogenized, so the top of each bottle sported a small layer of cream. As Bok set one of the bottles down in the kitchen, he studied the tiny globules of fat floating around the cream layer. "My God!" he suddenly realized. "These look just like my globules!"

Other members of the astronomical community didn't accept Bok's idea that globules represented an early stage in stellar evolution. One was Walter Baade from Palomar Mountain. Baade, seemingly alarmed, wrote to Bok with an alarming counterproposal that these dark spots were artificial inky spots involved with the finishing process of the photographic plate. "I am afraid," he wrote, "that you have been misled."[12] Baade's letter sent Bok into a frenzy of examining far more detailed photographs of areas rich in globules, like the Lagoon Nebula (Messier 8) in Sagittarius. "From the material available to me," Bok replied, "there is little doubt about the reality."[13] George Herbig, a specialist in star formation, also had his doubts. He thought that the globules might be transitory wisps of gas that appear and disappear. But then Bok challenged his friend to look at one in a telescope, rather than just on a photographic plate. "They seem as physically real as a globular cluster," he insisted. Herbig followed Bok's advice, and instantly became a convert.

The matter rested there, somewhat inconclusively, for the next eight years. In 1956 the new Palomar Observatory Sky Survey, taken with the beautiful new forty-eight-inch Schmidt camera, provided an answer. A search of two of its photographic plates revealed 17,000 globules. These more distant ones averaged only a thirtieth of the diameter of the Moon. At the same time, radio telescopes also started examining the globules, and detected their extremely cold hydrogen and carbon monoxide.

The globules were now accepted as a distinct and important class of objects by the astronomical community, and by the end of the decade they became publicly famous as well. Fred Hoyle, a prominent British astronomer, was probably the first to call them "Bok globules"—their current name—when he penned a science fiction tale called *The Black Cloud* that begins when astronomers using Palomar Observatory's eighteen-inch Schmidt camera (the same telescope the Shoemakers and I used to discover thirteen comets) found a dark cloud in space some five times the apparent diameter of the Moon. The object, it turns out, was "a fine example of a Bok globule."[14] As the astronomers continued to photograph the globule, they wondered how anyone could possibly have missed it on earlier photographs. They soon learned why: The globule was approaching the Earth with great speed, and finally one morning one of the observers rushed in, yelling "It's not there, sir! It's not there!" What, the other observer wished to know, was it that wasn't there? "The day, sir! There's no Sun!"

They rushed outside, and saw an amazing sight. "It was pitch black, unrelieved even by starlight, which was unable to penetrate the thick cloud cover. An unreasoning primitive fear seemed to be abroad. The light of the world had gone."[15]

Until this point in the narrative, Hoyle had correctly described a Bok globule, and its possible effect on the Earth should one happen to swallow us up. But as the story goes in a different direction, this particular globule turns out to be alive and intelligent. Just as the people in Hoyle's story were shocked to find that Bok had discovered not just dark globules but the first evidence of interstellar life, the globule was just as shocked to find intelligent beings living on a world instead of floating in deep space. The globule stayed for a few months, and after a long series of negotiations, it left the Earth in peace.

THE PILLARS OF CREATION

Although Bok did not discover these globules in the sky—Barnard found the first ones—Bok was the first to point out their astrophysical importance. He repeatedly emphasized that point: "The only thing that B. J. Bok did was to kick everybody in the pants and say you'd better pay attention to these Barnard objects. . . . In 1947 and 1948, Edith Reilly and I suggested the globules and no one believed that they were real, physical things. No one paid any attention until the radio astronomers came in."[16] Bok spent the rest of his life observing, studying, writing about, and lecturing on the Milky Way and its structures. He died in 1983, but if he were still around in 1995, he would have been amazed at a photograph snapped by the Hubble Space Telescope. It showed the magnificent dark nebulosity at the center of the Eagle Nebula, Messier 16. The pillars of darkness inside the Eagle are exquisite, one of the most stunning views ever seen with a telescope. The discovery team, led by Arizona State University's Jeff Hester, called these dark nebulae EGGs, for Evaporating Gaseous Globules. These globules are strongly related to Bok globules, but since they are some ten times smaller than the smallest Boks, they are not the same. "As a result," Hester explains, "the lifetimes of the EGGs to photoevaporation are fairly short as compared with most objects normally referred to as Bok Globules."[17] Moreover, while Bok globules are seen throughout the nebulae, these smaller globules appear along the very edge of the hydrogen molecular clouds.

Bok would have loved these EGGs, perhaps a subset of his globules, and he also would have happily approved of the acronym. In reality each evaporating gaseous globule is a cosmic egg out of which a single star will someday hatch. The photographs that reveal these pillars of darkness were not taken in a way familiar to Bok. He would have stayed with the telescope all night long, exposing the slow film until it had gathered sufficient light. Hester's team was nowhere near their telescope when it took its short exposures, for the Hubble Space Telescope orbits the Earth far out in space. But wherever one's telescope is posted, the spirit of the advice that Bok always gave his students would somehow still apply: Leave your telescope for a while, and just look up, "look up at the sky—just to make sure you are making bloody sense."[18]

Bok knew all about the importance of observing, and although he never lived to see it in space, he was well aware of the vast potential of the space telescope. In 1979, more than a decade before it was finally launched, he wrote an article called "The Promise of the Space Telescope," which appeared in the *Congressional Record*.[19] How remarkable it is that the telescope about which he was so enthusiastic performed as well as he predicted it would, and that one of its most dramatic images conveyed the magic of the globules he loved.

SIXTEEN

1951:
THE MILKY WAY
IS A SPIRAL GALAXY

Modern radio telescopes are exquisitely sensitive; a distant quasar is so faint that its detected radiation amounts perhaps to a quadrillionth of a watt. The total amount of energy from outside the solar system ever received by all the radio telescopes on the planet Earth is less than the energy of a single snowflake striking the ground.

—Carl Sagan, 1980[1]

One of the seminal discoveries of modern astronomy is the existence of the most listened-to radio station in the galaxy—a station so popular it goes by only a single call letter—H. It is a real radio station, and can be heard on any radio capable of picking up faint signals at 1420 megacycles, or 21 centimeters. The station offers a single tune—the natural broadcast of hydrogen atoms.

The story of how we found this station, and put it to use, began late in the 1930s, when Grote Reber, an amateur astronomer from the town of Wheaton, Illinois, submitted an article to the prestigious *Astrophysical Journal*. The young man's subject was rather unusual—almost unheard of, in fact—and concerned using telescopes to listen to space, rather than to observe it. Only one other person, Karl Jansky, had ever tried to hear the stars. Now a second observer not only was doing radio astronomy, but he was also suggesting the novel idea that he discovered radio radiation coming from the center of the Milky Way.

So unusual was the proposal that the journal's editor, Otto Struve, and the paper's reviewer, Bart Bok, decided to take the time and trouble to visit Reber's home in Wheaton. The visit was enlightening. The astronomers viewed the large collection of wires and rigging, and heard a comment by Reber's mother, who complained that the contraption, big enough to be visible throughout the neighborhood, was a nuisance that interfered with hanging up the wash on the family clothesline. At the time, Bok and Struve had no way of evaluating Reber's claim about the radio energy he claimed to hear from the Milky Way. Bok thought that the young man deserved a chance and that this new type of research looked promising. Accordingly, Reber's paper, entitled "Cosmic Static," appeared in 1940.[2] That paper set in motion an extraordinary flood of events that, within fifteen years, resulted in giant radio telescopes sprouting all over the world. Reber would be famous as one of the founders of radio astronomy.

OTHER GALAXIES TEACH US ABOUT OUR OWN

Shortly after Reber published his article, Walter Baade, a German astronomer at Mt. Wilson Observatory, was in a desperate situation. Baade had been admitted to the United States some years earlier, but when the U.S. entered the war against Germany in 1941, Baade, along with all other Germans in the country, was ordered to produce his immigration papers. Despite efforts by the other astronomers at Mt. Wilson to encourage their colleague to keep his papers in order, Baade had lost them. When immigration officials interrogated Baade, they were suspicious of his story, as they were not likely to trust the tale of an undocumented German citizen during a time of war. Considered an enemy alien, Baade was ordered to be placed under house arrest. However, with the help of the observatory senior staff, Baade was able to snatch an opportunity out of the disaster. He was permitted to serve out this house arrest, which would last throughout the war, among the world's largest telescopes at Mt. Wilson Observatory.

Since the observatory's other astronomers were absent from their telescopes to work instead on war-related projects, the mighty 100-inch telescope was available to Baade for almost four years. Added to this, the wartime blackout of the lights of nearby Los Angeles made the Mt. Wilson sky the best it had been in decades. Baade took great advantage of the rare situation in which he found himself. He concentrated on following the work Hubble had started, photographing the spiral patterns of galaxies as an adjunct of Hubble's work on redshifts. While studying these spiral patterns, he discovered that the stars in the spiral arms were generally of different spectral types from those of the stars in the galaxy's central bulge. Accordingly, Baade defined two different populations of stars. His "Population I" stars included the giant blue suns of spectral class O and B; Baade discovered that this type of star is common in a galaxy's spiral arms. (Henrietta Leavitt's colleague Annie Jump Cannon had organized stars by their spectral classes O, B, A, F, G, K, and M.) Moreover, he found that large clouds of hydrogen molecules also turned up in the spiral arms. The hydrogen nebulosity tended to accompany the blue stars as halos of nebulosity. Consisting of ionized hydrogen, the halos measured from 80 to 250 light years across, and were centered on the star.

The centers of the galaxies Baade studied, on the other hand, comprised the stars of Population II, a giant group of stars relatively devoid of both hot stars and hydrogen clouds. Baade believed that finding regions of Population I stars could be a way of tracing spiral arms in our own galaxy. In 1949, Baade wrote the following prediction in a letter to Bart Bok about variable stars in the constellation of Cygnus. As we have already seen, Bok was busy at the time with the globules that represented stars in the process of formation, but he was also seriously interested in studies of spiral structure. "From my studies of the Andromeda Nebula," Baade wrote, "I would bet the absorption for the 4 Cygnus Cepheids is due to the fact that the line of sight (from them to us) runs in the absorption-free—or at least absorption-poor, space between two neighboring spiral arms."[3] The line of sight, Baade meant, was unhindered by intervening layers of dust that would cause the stars to appear redder than they really are.

Bok tried to follow up on Baade's work with his own statistical studies of count after count of stars. However, by 1951 his observations were inconclusive on the great question of the true structure of our galaxy. The idea that our galaxy is rotating in space like a wheel helped spur three scientists from Yerkes Observatory—William Morgan, Stewart Sharpless, and Donald Osterbrock—to suspect that they could determine spiral structure in the Milky Way. That year, Morgan's group completed its preliminary studies of the distribution of stars of Population I and Population II. Their study was an attempt to determine groupings of stars in the spiral arms by measuring the distance to us from each blue star. They did this by measuring the spectra of the stars and plotting the results on a Hertzsprung-Russell diagram. By comparing each star's apparent brightness with its actual brightness as determined by the diagram, they were able to determine the star's approximate distance. Within a few months, enough distances had been worked out to paint a picture of two well-defined concentrations. Each one formed a long, narrow band, and the two bands were separated by a dark space. Comparison of these bands with photographs of the spiral arms in the Andromeda Galaxy showed conclusively that these bands are segments of two spiral arms in our own galaxy.

The newly found structures were called the Orion and Perseus arms, plus there was a portion of a third called the Sagittarius arm. The Orion and Perseus segments are each about 10,000 light years long, are parallel to one another, and are about 7,000 light years apart. Our own Sun, Morgan's team went on, lies near the inner edge of the Orion arm, the arm closer to the galactic center.[4]

Morgan was ready to announce his results at the Cleveland, Ohio, meeting of the American Astronomical Society, held in 1952. The night before he read his seminal paper, Morgan asked Bart Bok and his wife, Priscilla, to meet with him in his dormitory room. Sitting on the bed, the Boks looked on amazed as Morgan showed them the data that showed evidence for the three arms. After Morgan's presentation the following day, the room erupted in sustained applause: The idea was original and important, the research neat and straightforward, and the conclusion profound.

THE EMERGENCE OF RADIO ASTRONOMY

The work that Morgan started was about to get a huge boost from the new quarter of radio astronomy. Since the giant radio telescopes that were sprouting up in England, the United States, and Australia "see" a different wavelength of sky, through them, the Milky Way's spiral shape could be mapped far more effectively than by optical means. Morgan's team did work that unlocked the door; the radio telescopes allowed astronomers to enter the galactic room.

Radio telescopes are large dishes that receive radio waves. In just a few years, radio telescopes had evolved from tiny backyard experiments to giant dishes capable of detecting tiny amounts of radio waves at great distances. In 1944, just four years after Reber's paper, H. C. van de Hulst proposed that radio signals should be able to detect the hydrogen signature in the galaxy. The reason for his idea: hydrogen has a single, positively charged proton in its nucleus. Circling it is a negatively charged electron. The electron orbits the proton, but it can suddenly change orbits. If it jumps to a higher orbit, it absorbs energy, but if it closes in on the proton it releases energy. This change in energy level is fixed and is measurable.

Some of these changes resulted in visible glows, like the bright glow we see when we look at the Orion nebula. Others are more subtle. If the electron is in the closest possible orbit to the nucleus, a situation called the ground state, it can split into two slightly different energy levels called hyperfine levels. The change between these levels can be "heard" by radio telescopes at 1420 megacycles, or 21 centimeters. In the extreme cold of space, hydrogen atoms should exist at their ground states, so the 21cm band is where they would be detected.

In 1951, using a small pyramid-shaped horn antenna mounted on a roof of Harvard's Civics Building,[5] Harvard physicists Harold I. Ewen and Edward M. Purcell detected the galactic radio station of radiation from neutral hydrogen atoms at 21cm as a radio signal from the Milky Way.[6] Building on this discovery, Jan Oort was able to detect where hydrogen gas lies in our galaxy. The spiral arms are traceable by observing where hydrogen is especially concentrated. Not only could the Orion and Perseus arms be confirmed, but the arms could also be extended much farther out, beyond the dark matter that blocks the view of the optical telescopes.

Gauging the distances to these clouds was another matter entirely. By observing the spectrum of a hydrogen cloud through an optical telescope, Oort could tell from its redshift whether the cloud was moving away from or toward us. What Oort detected was amazing—different clouds in different places with similar spectral shifts meant the presence of a single arm stretching across the sky. Oort also found, in the same place in the sky, evidence of two different shifts, inferring the presence of two different spiral arms. Oort also confirmed the Sun's position in the Orion arm, the same structure that contains the stars of Orion with its majestic nebula. Moreover, he confirmed that the Perseus arm is some 7,000 light years farther from the center of the galaxy than we are. The Perseus arm contains the incomparable Double Cluster, one of the most spectacular sights in the northern sky. That arm has been traced almost half way around the galaxy.

The discovery of spiral structure was a highly complicated episode in astronomy that involved groups of people, from the engineers who built the giant radio dishes to the optical and radio astronomers who made the discoveries. But the work of the giant radio dishes had just begun; by the 1960s these telescopes would be opening a new chapter of discovery that would extend to the very edge of the universe.

SEVENTEEN

1963:
SEYFERT GALAXIES
AND QUASARS

I was in a complete state of shock.
—Martin Schmidt, 1963, on realizing that
3C 273 was 1.5 billion light years away

The story of the great quasars, it might be said, began late one 1930s evening in Cleveland. Bart Bok was lecturing there about the Milky Way galaxy, and in the audience was an enthusiastic young amateur astronomer named Carl Seyfert. They talked briefly after that lecture, and Bok was delighted when Seyfert later came to Harvard and enrolled in his introductory astronomy course. At graduation Seyfert was chosen to present the Ivy Oration, a valedictory address. Thankfully, Seyfert showed Bok his speech in advance. Writing of the high prices that some early astronomers paid for their beliefs, Seyfert claimed proudly that Copernicus did not die in vain at the stake. "It was Giordano Bruno who did not die in vain at the stake," Bok corrected his student regarding Galileo's contemporary who was burned for heretical beliefs in 1600; "Copernicus died peacefully in bed half a century earlier with his new book, *De Revolutionibus*, in front of him."[1]

Seyfert went on to Mt. Wilson, where, around 1943, astronomer Rudolf Minkowski told him that he had obtained some strange spectra, containing bright emission lines, of galaxies with bright, almost starlike centers. Seyfert took up Minkowski's suggestion and investigated this type of galaxy. These galaxies, of which Messier 77 is the best known example, are great spiral systems with very bright centers. After some two years at Mt. Wilson Seyfert went to Vanderbilt

University in Nashville, where he built the now-prominent Arthur Dyer Observatory. In the late 1950s a local radio station asked him to become their late afternoon weatherman. Seyfert accepted, and soon he was as popular a weatherman as he was a noted astronomer. "He tried to do so much," Bok said. "He would work at the observatory and then race over to the radio station." In June 1960, while driving to the station to deliver his forecast, Seyfert was killed in a car accident.

THE THIRD CAMBRIDGE CATALOGUE

By the end of the 1950s, radio telescopes were making so many interesting discoveries that astronomers were arranging to have them built in many countries. When the Cambridge Radio Astronomy Observatory was completed not far from the university rugby ground, an early goal for this observatory was to prepare a catalogue of radio sources in the sky. Since the resolution of a single radio telescope is far less than that of an optical telescope of the same size, accurate positions of radio sources were hard to determine. That changed somewhat in 1953, when Cambridge opened a different kind of radio telescope consisting of a series of small antennae, widely spaced from one another, and used as an interferometer. The technique of interferometry uses separated telescopes that are linked together electronically; the linked telescopes have a resolving power equal to the distance between the two farthest telescopes. With this increased resolution, astronomers could use radio telescopes to derive positions of astronomical objects with almost the accuracy of optical telescopes.

With this new ability, Cambridge began work on lists of radio sources. The "Third Cambridge" radio survey of 470 radio sources was published in 1959 and revised two years later. Astronomers would soon find some of the most the dramatic objects ever observed in nature in that simple list. In 1960, Allan Sandage used the catalogue position and the 200-inch telescope at Palomar to identify the optical "counterpart" of one of the Cambridge radio objects. Through the radio telescope it was number 48 in the third Cambridge Catalogue, or 3C 48; through the power of the mighty 200-inch, it was a faint star. Next, Sandage's colleague Jesse Greenstein analyzed the spectrum of this star. He found it unusually rich in the ultraviolet, unlike anything he had ever observed. 3C 48 was a radio star, or a star with very powerful radio emissions. It was not likely to be nearby.

With 3C 48 proving to be such a mystery, it was time to identify and study the optical counterparts of other 3C radio sources. Cyril Hazard, a radio astronomer with the Parkes Radio Telescope in Australia, had an ingenious idea for determining a position of a radio source with better accuracy than any radio telescope could muster. Object 3C 273, in Virgo, happens to be on the ecliptic—the plane of the solar system—which opened the possibility that the Moon might occult it from time to time. Since the Moon slides way north or south of the ecliptic during its monthly swing around the Earth, it might, for a time, occult certain stars on a particular band of the ecliptic, then inch away from that band and not occult those stars for a period of many years. It was one of these happy coincidences of fate that just when it was needed, 3C 273 happened to lie on the path that the Moon would soon cover.

As the Moon moved in its eastward orbit, it passed directly in front of 3C 273 not once but three times in just a few months in the early 1960s. The instant the Moon passed in front of the radio source, the far-off object's radio signal was cut off abruptly, and then about an hour later, the signal reappeared just as suddenly. The only problem with the observation was that from Australia, the Moon was fairly low in the sky at the time of one of the occultations, so the telescope had to be modified somewhat so that it could point closer to the horizon than its design ordinarily permitted. The adjustment was worth it. Because we know the position of the Moon precisely at a given time, these events allowed astronomers to determine the exact position of the radio source. Armed with that information, Maarten Schmidt, a colleague of Greenstein's at Cal Tech, immediately found the star on the photographic plates that had earlier been taken at the observatory. He found a thirteenth-magnitude star, bright enough that I have since seen it through my own six-inch telescope. Then Schmidt turned the 200-inch on it. Through its great resolving power, 3C 273 looked like a round, starlike object that differed from a normal star in that it was clearly not a point of light—its edges were fuzzy. More important, a jet of faint light hung out from one side. When the radio observations were compared with the optical ones, it turned out that even though the jet was much fainter than the star, some 90 percent of the system's radio energy was associated with it.

Meanwhile, Greenstein and Schmidt were carefully examining the spectra of these objects. Greenstein thought they might be the remains of stellar explosions in our own galaxy. When Schmidt obtained a spectrum of 3C 273, he was astounded by how unusual it was. After weeks of study, he focused on three hydrogen-like lines that he would normally expect to see on the blue side of the spectrum. Instead, he was seeing these lines at the red end.

Then a stunning idea hit him: Could an extremely high redshift, where the entire spectrum is shifted strongly toward the red end, explain the spectrum? Puzzled that the H-alpha line, the strongest of the hydrogen spectral lines, was nowhere to be seen on the visible-light spectrum of 3C 273, Schmidt wondered if a device capable of identifying lines in the far infrared would detect it. The line indeed turned up at the extreme red end of the spectrum, and Schmidt's theory was confirmed: according to Hubble's law, 3C 273 is *1.5 billion light years* away; Schmidt was in "a complete state of shock" at this development.[2] Now abandoning his earlier "local" explanation of 3C 48, Greenstein found the same type of redshift and determnined that this object was 3.6 billion light years distant. If its redshift were interpreted correctly, 3C 273 is receding from us at almost a seventh of the speed of light, and 3C 48 is rushing away from us at more than twice that velocity.

Radio astronomers called these fantastic objects quasi-stellar radio sources; the optical astronomers called them quasi-stellar objects or QSOs, but the term "quasar" soon became their accepted moniker. A quasar is an extreme version of a Seyfert galaxy—the highly energetic core of a distant, in most cases elliptical, galaxy. Whereas the cores of Seyfert galaxies are several times brighter than the surrounding disk and arms of the galaxy, quasars are thousands of times brighter than the rest of the galaxies they inhabit. If the quasars we see were ordinary galaxies, like Andromeda, we wouldn't detect them at all because of their great distances. But if Andromeda Galaxy had a quasar at its center, it might blaze as brightly as Jupiter in our sky.

A quasar is the highly energetic core of a galaxy, and the only "machine" that can power something that powerful is a black hole. It probably formed by the collapse of a single star near the middle of the galaxy, probably in the galaxy's youth. Because stars are so close together in a galaxy's core, such a black hole could grow simply by attracting material from surrounding stars and gas. As gas approaches such a black hole, it settles into a disc that is dragged inward. At the inner edge of this accretion disc, the gas suddenly becomes very hot, emits gamma rays and pairs of particles and antiparticles. As these pairs of particles and antiparticles fall into the black hole, they add to its mass and spinning. In this way, the black hole swallows the equivalent of several times the mass of the Sun and converts much of it to energy *every year*. The supermassive black hole swiftly pushes electrons out along magnetic fields to speeds approaching that of light; the result is a form of radiation called synchrotron radiation. Surrounding it are rich clouds of gas that have erupted out of the center in times past.

If a quasar stayed that active throughout the age of the universe, it would quickly swallow its entire galaxy. It is possible that a quasar forms as a result of two colliding galaxies. The additional material at the center would provide the needed mass. Some evidence for the collision view was provided in 1983 by the Infrared Astronomical Satellite, which discovered many ultraluminous infrared galaxies (ULIGs) which appear to be pairs of colliding galaxies in which quasars are forming. We don't see these galaxies or quasars visually because they are still surrounded by dust. Within the ULIGs, we might be seeing how quasars are formed. After the collision, the new galaxy is elliptical and boasts a quasar.[3]

SEYFERTS AND BL LACERTAE OBJECTS

With quasars finally added to the growing list of unicorns in the universal zoo, by the end of the 1960s the Seyfert galaxies had become a major link between ordinary galaxies and quasar-centered galaxies. Seyfert had died with no idea of the important role his galaxies would play; in fact, his obituary in *Sky & Telescope* evaluated his rich career without specifically mentioning the Seyfert galaxies.[4] Later that decade, Bart Bok helped arrange a symposium on these galaxies, a sort of "send-off" for them, as Bok described it. During the symposium Bok formally suggested that the galaxy class be named both for Seyfert and for Minkowski, who had initially suggested this line of research to Seyfert. "Rudolf," Bok proposed to Minkowski, "it is about time the world knows that your name ought to be associated with these galaxies. They ought to be called the Minkowski-Seyfert galaxies."

Minkowski returned a puzzled stare to Bok and didn't say a word. "It was you who supervised Seyfert's work there," Bok added. Since he always tried to keep in touch with what his students were doing, Bok was well aware of how Seyfert's research started. But Minkowski would have none of it. "I don't remember a thing about this," he replied in front of the assembled scientists. The result: Seyfert's name alone is attached to these galaxies.

The quasar discovery story has yet another twist to it. In 1929 the Sonneberg Observatory in Germany reported a star changing slowly and irregularly in brightness. Since it was in the constellation of Lacerta, it was named BL Lacertae according to the rules for naming variable stars. The matter rested there for thirty-nine years until 1968, when a routine University of Illinois radio survey uncovered an object with strong radio emission that corresponded to the position of the variable star. Then a team of Canadian astronomers linked the radio source with the star. The real surprise came when astronomers recorded a spectrum which turned out to be utterly featureless, revealing neither the typical emission nor absorption lines that astronomers expect to see in stellar spectra. The second surprise was that the "star's" radio emission varied dramatically in periods as short as a few days—faster than any known radio source outside the solar system. Optically, variations were recorded by Canadian astronomer Rene Racine in as short a time span as a few minutes.[5]

By 1973, long exposure photography was showing that BL Lacertae is surrounded by a thin haze. Could it be the center of a distant elliptical galaxy? The answer came when astronomers finally had the detection equipment capable of recording a good spectrum of this extremely faint haze. The spectrum was typical of a galaxy, but not an ordinary one—this one had an extreme redshift that indicated its distance as a billion light years. We now understand it to be the very bright center of a distant elliptical galaxy.

An observational difference between quasars and BL Lacertae objects (some now call them blazars) is that although the cores of both must contain hundreds of millions of solar masses, the BL Lacertae objects vary more rapidly. The speed of this variation tells us something important: The BL Lacertae objects cannot be seen to vary faster than the time it takes light to travel across its diameter. This means that these very bright objects cannot be larger than our own solar system, and may be not much larger in space than the distance between the Sun and Earth!

As all these observations continued to come in, a picture was being built of the distant universe, filled with galaxies in which supermassive black holes had taken over a small fraction of their centers. But others questioned their interpretation of all this evidence. Could quasars be just weird objects in our own galaxy? Just as radio astronomy helped open quasars as a field of study, a new technology, CCDs, was on the horizon to provide an answer.

EIGHTEEN

1979:
USING A GALAXY
AS A TELESCOPE

Gravitational Lenses and Einstein Rings

> *It looks like God cut up a piece of rope and just plopped it in the sky.*
> —Vahe Petrosian, 1988[1]

When Galileo began the astronomical story of the telescope by looking at Jupiter through his in 1610 and discovering four moons, he could have no idea of the strange course that telescopes would take over the next 400 years. And he certainly would have no clue that the universe itself would help by providing telescopes of its own by using galaxies as lenses.

This chapter tells the story of how the remarkable development of gravitational lensing has helped change our view of the universe. It is a story that begins in the Big Dipper, and for me, a tour taken in September 1979 at the brand new Multiple Mirror Telescope (MMT). The tour took place just a few days after I moved my possessions and life to the astronomically charmed region around Tucson, Arizona, and arranged for a tour of the new MMT; Dan Brocious, the public information officer, obliged me.

As we approached the summit of Mt. Hopkins, I saw a most unusual telescope building—not a dome but a multi-story box, inside of which was a telescope composed of six seventy-two-inch mirrors. I also saw a tiny electronic chip, no larger than a 35mm film. "This is the new way of astronomy," Dan exulted, showing me the chip, a duplicate of which was actually at the common focal point of the six telescopes. "This is a charge-coupled device, or a CCD chip." The MMT, it turned out, was one of the first telescopes to use this new technology. Far more sensitive than film, CCDs will make any telescope much

162

more powerful than it could be with film, and the biggest telescopes, like this one, could uncover new astronomical territory undreamed of in the past.

This remarkable telescope with its CCD technology was barely out of the starting gate when I first saw it. Its official opening had taken place only a few months earlier. What I did not expect to see that visit was evidence that it had already made a major contribution to astronomy. With great excitement, Dan showed me one of its first images. Taken a few weeks earlier by a team including Nathaniel Carleton and Fred Chaffee, using the MMT, the image clearly showed two quasars only 5.7 seconds of arc apart. Together the quasars are named by their coordinates in the sky, 0957+561.

First detected by a radio survey using the giant Jodrell Bank telescope in England, the quasars' initial position in the sky was inaccurate until the 300-foot long antenna at Green Bank, West Virginia, obtained a more accurate fix. With a better position, observers were able to find the twin objects on the Palomar Sky Survey. The first spectrum showed that the two objects were practically identical. Both were racing from Earth together, at identical velocities. The chances that two real quasars would have such identical stories were very low. Perhaps some intervening galaxy might be acting as a lens, splitting the light so that it looks like two quasars instead of one.[2]

HOW TO MAKE A GRAVITATIONAL LENS

The idea of gravitational lensing is actually quite easy to explain using the broken stem of a wine glass. Hold the stem so that the base is away from you, point it toward a light, then watch how the base acts as a lens. Depending on how you hold the stem, the base will divide the light into two or more separate lights. Or, if you hold the base exactly right, there will be a ring of light around the base.

What Dan Brocious was showing me that afternoon was the universe's version of the stem of the wineglass. It was the first image showing both the two quasars and a faint hazy light slightly closer to one of the quasars than it was toward the other. Here was the galaxy, a fraction of the distance to the remote quasar, *acting as a lens* that was splitting the light of a single quasar, making us see it as two quasars! Ecstatic astronomers were watching the galaxy and the quasar playing out an important part of Einstein's General Theory of Relativity— that light bends when passing near a strong gravitational source. In Einstein's own lifetime, the bending of light by the gravitational field of the Sun during an eclipse was detected (see chapter 13). A quarter-century after Einstein's death, manifestations of his theory are showing up far afield in the universe.

Because of the effect of the lens, it is likely that these two quasars' images are quite a bit brighter than they actually are, since the gravitational lens caused by the galaxy is magnifying their images. It is interesting to note Nature's irony here: The intervening galaxy is too faint to detect except by using the most powerful telescopes, and yet the objects it allows us to see become much brighter. The galaxy is a lens, a catalyst that allows us to detect what lies beyond it.

GALACTIC ARCS

By 1987, CCD technology, with its ability to peer ever deeper into the cosmos, was revealing even more exotic structures. In January of that year Roger Lynds of Kitt Peak National Observatory and Vahe Petrosian of Stanford announced their discovery of two giant arcs of light surrounding the distant clusters of galaxies Abell 370 in the constellation of Cetus and 2242-02 in Aquarius. A second group led by Geneviève Soucail recorded them at almost the same time using the Canada-France-Hawaii telescope at the summit of Mauna Kea. These arcs, which look like ropes hanging in the sky, are truly enormous, at least 300,000 light years long, and have the light output of 300 billion suns.

What could generate that much energy and then spread it all in an arc? Interpreting the arcs' simple structure was a problem from the start: "A nice simple structure like that," Petrosian noted, "gives physicists nightmares."[3] These discoveries were in turn based on observations by Lynds and Petrosian taken as early as 1976 on the galaxy cluster Abell 2218, but those pre-CCD images were not good enough to permit detailed study.

If the arcs are real structures, how can they stay together over such a large distance? Possible explanations were these: If a wandering galaxy had just fallen into the cluster, a long streamer might result, or it could be a jet erupting out of a colossal black hole at the center of the cluster, in the same way a jet comes out of Messier 87. But this arc is too large to satisfy those explanations. "We are dealing," Petrosian said, "with a very unusual and potentially important phenomenon."[4]

A few months later, a new interpretation arose: Could the arc be a light echo from an ancient eruption of a quasar? When the quasar flared, a spherical "echo" of light would then travel outward through space.[5]

During this time, Geneviève Soucail focused her attention on a different possibility: that the arcs were the result of a gravitational lens at work. This would certainly help explain the sheer size. If the arc really did contain the equivalent of 300 billion suns, it would be the size of a galaxy. Perhaps that is exactly what's happening: The arc is a galaxy whose light—at least as far as we on Earth are concerned—is spread over an enormous distance by gravitational lensing.

By the summer of 1988, the mystery of the arcs was leaning strongly in favor of Soucail's explanation. When we view these arcs, we really see an image of a remote galaxy stretched out into an arc by the gravitational field of an intervening cluster of galaxies. The cluster acts as the objective lens of a gigantic telescope, with the actual telescope, through which we image the arc, being the eyepiece. The telescope is not an optically good one, since the galaxy's image is smeared over a wide area as if someone spread jelly on the lens. However, were it not for the intervening cluster, we would not know of the distant galaxy's existence at all.

FINDING MORE LENSES

With the theory of gravitational lensing rapidly gaining favor, the next step would seem to involve getting better images of other galaxy clusters in the hope that they, too, would show the smeared out images of far more remote galaxies. If the gravitational lens is precisely on a line of sight from the very distant galaxy to us, the optical effect would make the galaxy appear as a ring around the lens. As Petrosian said late in 1988, it would be spectacular when "someday, somebody finds a perfect ring."[6]

Petrosian didn't have long to wait. Ian Browne led a group that used the Very Large Array Radio Telescope in New Mexico to image B0218+357 in the constellation of Triangulum. The radio telescope "saw" two quasars, each varying in brightness in the same manner, but not exactly at the same time, because the light paths from the quasars are slightly different. However, one of the images is surrounded by a perfect ring which appears to be the smudged-out image of a jet bursting from the center of the quasar.

This new field of lensing is opening another important area of research. From the time delay of the variations in double quasars, astronomers can infer the distance of the intervening galaxy; that is, the galaxy doing the lensing. In other words, gravitational lenses could provide a yardstick for measuring distances in the universe, just as Cepheids and redshifts did in earlier generations.

Gravitational lensing is indeed a strange new branch of astronomy, a field totally unheard of a generation ago. It is allowing us to measure the spectra of galaxies that we would have not the slightest hope of detecting were it not for the auspicious presence of a lens in the form of a galaxy or cluster of galaxies. When we take the spectrum of an object we need to see only its light, not a beautiful, unsmeared image. We also learn something about the galaxy or cluster doing the lensing. When we study the geometry of how a galactic arc is spread out, we can learn about how mass is distributed in the lensing cluster.[7] In 1994, for example, Gerard Luppino and colleagues imaged a cluster of galaxies in Taurus, MS 0440+0204. In the center of the cluster lie several galaxies so close together that, over the next billion years or so, they will probably collide with one another and merge into a giant elliptical galaxy. Surrounding the cluster is a network of faint blue arcs. In addition to being interesting by itself, the cluster is a gravitational lens.[8]

Astronomers have imaged large numbers of double, triple, and quadruple images as well as arcs. Einstein's cross, also called Huchra's lens, is one of these. The lensing galaxy is so elegantly positioned that the quasar so far behind it is broken up into four images that resemble a sort of cross. As astronomers probe ever larger concentrations of mass to find more powerful lenses, finding arcs should prove easier in the future.

THE CONCEPT OF DISCOVERY TEAMS

Unlike other discovery stories in this book, the tale of gravitational lenses cannot focus on a single person. Virtually every discovery image, and every competing theory, involved teams of astronomers. The whole story, in fact, would be impossible if the telescopes themselves weren't teamed up; both radio and optical telescopes were needed to discover these lenses. This is the direction that astronomy, like other sciences, has taken in the last fifty years.

Certainly a discovery as important as that in 1610 by Galileo, in our age, would have been made and published with a long series of well-known names, followed by the lesser known names of students and assistants. This point was brought home to me last spring when I attended the award ceremony for a science contest in which children across the United States were asked by *Parade* magazine and the National Science Foundation, (NSF), to arrange themselves in groups of four to come up with a science experiment. When I asked why a child couldn't submit an entry independently of a group, I was informed that since virtually all science was now done in teams, the NSF would not allow an entry by a child working alone.

Teamwork and structure are hallmarks of this brave new world of scientific discovery. Can magic, spirit, and imagination also be hallmarks of this new era? Will a lone Galileo ever again be able to cobble together an instrument, point it at the stars, and change our conception of the Universe?

NINETEEN

1993:
COMET SHOEMAKER-LEVY 9

Do Not Go Gentle into That Good Night

> *Old age should burn and rave at close of day;*
> *Rage, rage a gainst the dying of the light.*
>
> —Dylan Thomas[1]

The bewildering story of gravitational lenses brings us to the end of the twentieth century, and supposedly into a completely new type of observing that uses electronic chips to peer to the very edge of the universe. The story of communications through space also brings us to this new era of electronics. With this kind of discovery dominating the field, the age of the scientist calculating with slide rule and pencil, like the astronomer observing in the dome with a telescope, seemed to be over. But that old-fashioned era, though clearly in decline, was not quite gone yet as the century neared its end. The story of this next discovery brings us back to the simpler time, and back to our own solar system, our home.

Comets, I always thought, lasted forever, even though their visits are brief. Their filmy tails and gracious movements across the sky seemed to provide a sense of permanence. Then came Halley's Comet in 1986, and with it an astronomer's calculation that about one yard of cometary material is lost each time it spends a few months close to the Sun. During those few months, the Sun's heat causes the comet's ices to sublimate (turn directly from ices to gases), releasing gases and dust. Since Halley's comet is at least seven miles across, it would seem that there is an inexhaustible supply of activity for the comet.

We now know that another process is at work each time a comet rounds the

Sun. As the comet cools, a crust forms around it, making it slightly more difficult to sublimate the next time the comet comes around. After hundreds or thousands of returns, the crust layer might become so thick that the comet shuts down completely, unable to grow a coma and a tail, and unable to behave as a comet should. The result is that the comet has become extinct, in that it has no coma, no tail, and behaves like an asteroid.

COMETS AND THE MECHANISM OF LIFE

In the long run, then, comets are not forever, or at least they do not behave like comets forever. In 1994, we learned that lesson in a big way when Comet Shoemaker-Levy 9, in an amazing series of colossal explosions, simply incinerated itself in the clouds of Jupiter. Now we knew that a comet's death can be spectacular if it involves a collision with a planet. The concept of collisions is not an accidental, or even an unfortunate, part of the cosmic agenda. They are a necessary part of the mechanism of life.

When the Earth was young, it was far too hot to prevent vaporization of any organic molecules which might have been in our part of the primordial cloud from which the solar system was formed. In order for these organic molecules to become part of our planet, they needed to remain in relatively cool parts of the solar system and then, once the Earth cooled off, they needed somehow to get delivered back.

Comets provided an ideal transportation system to accomplish this. After the Earth cooled off some 4.5 billion years ago, a long period of heavy bombardment took place during which comets struck the Earth perhaps once a century or millennium. The bombardment went on for at least half a billion years, ending in a grand finale of strikes some 3.9 billion years ago. Some of the largest impact basins on the Moon, like the Imbrium Basin, were formed at that time.

This long series of comet impacts, aided by a constant wafting-down of comet particles, gave Earth much of its supply of carbon, hydrogen, oxygen, and nitrogen. After a long period, a number of things might have happened. In one scenario, simple organic materials, like hydrogen cyanide, condensed to form biomolecules like amino acids. It is possible that this process could have taken place on Earth, in comets, or even in the original solar nebula. We now know that comets contain substances as complex as hydrogen cyanide, and formaldehyde, for in 1986 the *Giotto* spacecraft detected a polymerized formaldehyde on Halley's Comet.

In step 2, clay minerals catalyzed to form the first nucleic acids, and during step 3, specific bases formed the first RNA molecule, helped along perhaps by the intermediate formation of metal ions. With step 4, RNA began to reproduce itself without the need for the metal ion catalyst. Finally, in step 5 several new enzymes were selected to form the first DNA molecule. Through DNA, life on Earth had a good start.[2] However, life might have had a hard time continuing. At any step in this possible process, the whole thing could be stopped dead with the crash of a new comet. But about 3.9 billion years ago, the bombardment halted. The environment of Earth quieted down. There would still be major comet or asteroid crashes—one, some 65 million years ago, probably led to the extinction of the dinosaurs and the rise of the mammals—but nowhere near as many as during that period of heavy bombardment.

As our own civilization grew, our understanding of the importance of comet impacts did not seem to change much. In the mid-nineteenth century, the British astronomer J. Russell Hind proposed that if a comet were to strike, it would "perhaps be comparable only as regards the mechanical effect upon the earth to a meeting with a huge cushion."[3]

PERSUADING THE PUBLIC
THAT COMETS POSE A HAZARD

In January 1993, I attended a meeting about hazards that could be caused from comets and asteroids. One of the biggest concerns that the attending scientists had was what to do about the "giggle factor." Whenever anyone suggested that comets could pose a hazard, the press responded with sarcasm. In those early days of 1993, it was difficult to make anyone take the threat seriously. The idea of the meeting was to produce a book that would help galvanize public attention to impacts.[4] How could anyone know at that meeting that three of its partipants would discover, within a few weeks, a comet that would spotlight the consequences of impacts very effectively? Gene and Carolyn Shoemaker were prominent participants at that meeting, and during its course they and I quietly made plans for our next few observing "runs" at Palomar Mountain.

PALOMAR, MARCH 1993

Just before our third observing run of that year, I set up my camera in my back yard and took a thirty-minute photograph of the Milky Way that I hoped to use for a biography of Bart Bok I was writing. The picture was successful, and eventually appeared both inside and on the cover of that book.[5] It had to be on its way to the publisher before I was on my way to Palomar Mountain, but by March 22, both picture and I were off to new locations.

That first observing night was spectacular. A storm had just passed through, leaving a pristine sky. With great enthusiasm the Shoemakers and I began taking eight-minute-long exposures, but that soon stopped when Gene Shoemaker realized that our supply of film had been damaged; in the few weeks since we were last at Palomar in February, someone had opened the box containing all the film we had carefully hypersensitized for six hours to increase its ability to record light efficiently. Without any more hypersensitized film, we were out of luck for that night.

Or so we thought. Gene tried an experiment: Thinking that since the six-inch-diameter circular films had been stored flat, one on top of the next, he hoped that the uppermost films might have protected the lower ones. He took out two films slightly beneath the topmost layer of films and quickly developed them. Gene was right: Except for two areas that had been badly exposed to light on each film, the centers of the films seemed light enough to accept the dark negative images of starlight. In this way, we managed to make our way through the next six hours as fresh films were being hypersensitized. By about 3 A.M., we rescued the new films, and for the next two hours were able to use perfect films. I fully expected that Gene would discard the rest of the light-struck films.

March 23, 1993

The next day was beautifully clear as we looked forward to a second full night of observing, this time with fresh, unexposed films. We began that night smoothly, but after two hours an ever-thickening layer of cirrus clouds moved in, the harbinger of a new storm. We knew that our observing time this month might be shortened by winter storms, but as the clouds swung quietly across the sky, we decided to stop right there. Down the stairs we walked, and stepped out the front door. Ever the optimist, I suggested that the sky wasn't really that bad. In response, Gene gave me a financial argument. Each time we slap a film into the telescope, he explained, it costs us four dollars. We couldn't afford to be taking pictures unless the sky is completely clear in the region we were searching through.

As I looked at the clouds, I thought of Gene's argument, and then a thought hit me. "Gene," I asked, "how many bad films do we have left from last night?"

"About a dozen. Why?"

"Those films," I went on, "won't cost us a cent. They're already wasted. Why can't we continue, even this fairly poor sky, with those?"

In the minute of silence that followed, Gene, Carolyn, and I looked at the sky, and at each other. The sky even seemed to be improving a bit. Finally Gene spoke. "Let's do it!" he said emphatically. We rushed inside, and I climbed the steps two at a time to the dome, while Gene loaded one of the filmholders with a partially useful film.

Once he was upstairs with the film and I had it loaded in the telescope, Gene read me the coordinates of the next field. I was astounded when I looked through the telescope; the area was so bright I could barely see the guide star. I looked through the slit in the dome to find out why, and realized that Jupiter was very close by, its light swamping everything. "Maybe we should skip this field," I suggested.

"Can you see the guide star?" Gene asked. When I answered that I could, he said that since this was indeed the next field, and since we were already there, we should simply proceed with it. And so, the night's comedy of errors went on. I completed the exposure, and the next two after that, before more clouds came by and we had to stop and wait. Ideally, Carolyn, who did the scanning, liked to have about forty-five minutes of time between films taken at the telescope; if much more than an hour separated the two films of a pair, it became more difficult for her to spot the floating effect of a new comet or asteroid.

On this evening, we waited well past the forty-five minutes. An hour passed, then an hour and a quarter, then an hour and a half. I kept on rushing outside to see if there was any change, but all I saw were thickening clouds. "We'll just have to deal with these as singletons," Carolyn quipped, using the term we have for fields for which only a single film is available; no motion can be detected, and it is very difficult to confirm a comet. After a little more than an hour and forty minutes since we began the first exposure of the Jupiter field, I went out again and noticed a small clearing approach the area around Jupiter. "If we're lucky," I reported, "we might be able at least to get the second film for the Jupiter field." I went upstairs, found the position of the guide star near Jupiter, and waited. Other than the rapidly moving clouds overhead, everything was perfectly calm beyond the opening of the dome. There was no wind, no other noise. Just quiet. Suddenly I glimpsed the guide star! The clear area was close now, and we had to begin immediately if we had any hope of completing the eight-minute exposure before the break in the clouds passed.

We were lucky. The clouds held off until the last thirty seconds or so of the exposure. We had no hope of completing the pairs of the other two fields we had taken earlier, but I was happy we had gone this far. When Gene developed the fields, he momentarily panicked when he saw a huge black spot on each of those fields, then he laughed when he realized that Jupiter, not exposed films, was the culprit this time.

March 25, 1993

On the following evening we took a handful of pictures before thick cirrus clouds ended our observing. The next day, March 25, became windy and snowy as the storm hit with all its force. But just in case it were to clear, we wanted to be prepared, so Gene and I hypersensitized a new batch of films. We returned to our observatory dome with the fresh film around 4:00 in the blustery afternoon. Gene sat down to read, I set to work on *The Quest for Comets*, a book I was writing at the time, and Carolyn inserted the two Jupiter films into her stereomicroscope.[6]

For the next ten minutes the dome was fairly quiet, with the hum of the stere-omicroscope and the clacking of my keyboard being the dominant sounds. We had a guest with us, Philippe Bendjoya, an astronomer from France. With things seem-ingly very slow, he left the dome for a while. As Carolyn continued to move the stage back and forth, occasionally we could hear a squeaking sound, but there really was no special noise when the stage was moving. So at first Gene and I didn't notice when Carolyn stopped its motion, moved it back a bit, then sat up straight in her chair and stared intently into the pair of eyepieces.

"I don't know what I've got," she told us, "but it looks like a squashed comet."

Gene and I then had our turns to witness the strangest looking comet we had ever seen. Not a single coma and tail, this maverick comprised multiple heads, each with a tail, and two pencil-thin lines on either side. With no possibility of our being able to rephotograph this field, I called my friend Jim Scotti, whom I knew was at that moment observing at the thirty-six-inch Spacewatch camera at Kitt Peak, some 400 miles east of us. The storm that clouded us out was on its way there, but he still had a hazy, though usable, sky. We asked him to check on our observation, which he suspected was a reflection of Jupiter off the optics of the telescope. Meanwhile, Gene and I, aided by Jean Mueller of the Palomar Obser-vatory Sky Survey, obtained accurate positions of the two images. Two hours later, the task accomplished, we returned and phoned Jim once again. "That sound you heard," Jim began, "was the sound of me trying to pick my jaw off the floor!"

"Do we have a comet?" I inquired.

"Boy, do you three *ever* have a comet!"

As I repeated Jim's words to Gene and Carolyn, we just stared at each other. After I hung up the phone, the "whoopee!" we let out could probably be heard all over the mountaintop. We called Jean Mueller, who rushed over to share the excitement. There were hugs to spare all around that night. It's important to remember that we had absolutely no idea of the importance this new comet would later assume. All we knew was that we had stumbled onto the most unusual looking comet any of the five of us had ever seen. Based on the hundred-plus comets that Carolyn, Gene, Jean, Jim, and I had discovered or studied over the years, it was safe to conclude that our new discovery was probably unique.

The next morning, we awoke with the giddy feeling that comes with having been out too late the night before. The next few nights were cloudy, so we had no opportunity to observe our comet. But other observers did; through large tele-scopes the comet sported no fewer than twenty-one fragments. On the last night of the run, the sky cleared around midnight for a brief time; we rushed to the dome and rephotographed the field. The comet was low in the sky, but it was still there! I'll never forget the high spirits I enjoyed during the 400-mile drive back to Tucson after that observing run.

MAY 22, 1993

Within the next few days, observers from all over the world would image the comet that eventually became known as Shoemaker-Levy 9, the ninth returning comet we had discovered as a team. Five other Shoemaker-Levy comets were eventually found with orbits that cause them to return to the Sun not at all or only after many hundreds or thousands of years. One of those comets, the Shoemaker-Levy then designated as 1993h, turned up on films we took two months after the discovery of Shoemaker-Levy 9. That comet made a nice pass through the southern sky, but did not become very bright. This visitor, 1993h, minded its business as it coursed through the solar system, unaware of the drama which on that very day was unfolding with its namesake.

A few hours earlier, while preparing for our night of observing, we learned through e-mail three momentous things about Comet Shoemaker-Levy 9. First, enough precise positions had come in to enable an accurate orbit to be calculated. Second, this comet was orbiting Jupiter, and eight months before we discovered it, the comet had approached Jupiter so closely that the planet's tidal force caused it to fall apart catastrophically. Third, the current orbit about Jupiter was to be its last, and in July 1994 the comet would collide with Jupiter.

As we read the announcement, Gene, Carolyn, and I were far more subdued than we were on discovery night two months earlier. Gene studied the orbital calculations while Carolyn and I asked him questions. In fact, the importance of the coming impacts didn't hit me until some days later, during a late-evening conversation with astronomer Clark Chapman and his wife, Lynda. In simple but compelling terms, Clark described their importance in a way Gene couldn't. This could be the most significant event in our solar system since the invention of the telescope, Clark explained. Something special was about to happen in our solar system, whether or not we actually witnessed any of it. In a year's time we would be treated to a lesson from Nature; in its dying moments, Comet Shoemaker-Levy 9 would show us how impacts shaped the solar system.

THE ONSLAUGHT BEGINS—
FOR THE COMET AND ITS DISCOVERERS

The next fourteen months were the most stressful in my life. As early as January 1994, CNN and *Time* magazine began planning their coverage with the Shoemakers and me. By spring Carl Sagan wrote me with some questions in preparation for an article he was writing for *Parade* magazine. In July alone I completed at least 150 separate interviews by virtually every major news organization, in TV, print, and radio; one was as ABC News Person of the Week; these interviews included groups ranging from the president and vice president of the United States to teachers and children.

Our observing runs throughout that year were interrupted almost daily by hordes of television cameras and reporters. We became the standing joke of the observatory staff—"here come the Shoemakers and Levy with their entourage!" At first unprepared for the onslaught, Palomar Observatory quickly became very helpful and cooperative both with us and with the various news organizations. The last observing run before the impacts ended early in the morning of July 15, 1994. I was already in New York, a guest on the *Today* show; Gene and Carolyn were being broadcast from Palomar. A day later, our team was reunited in the Space Telescope Science Institute.

RESOLUTION: JULY 1994

On the night of July 16, the grounds of the institute were a veritable forest of broadcast antennae. As Gene, Carolyn, and I led the first press conference by telling the story of the discovery of our now-famous comet, a team of scientists one flight below us was staring wide-eyed at the pictures coming in from the Hubble Space Telescope (HST). These pictures showed a plume rising 2,000 miles off the edge of Jupiter. There was a collective gasp in that Operation Support System (OSS) room. As the scientists realized what a gem they had captured through the electronic eye of the HST, one of the scientists, Heidi Hammel, broke the silence with a quiet "Oh, my God!"

Harold Weaver was another of the astronomers in that room. "We realized that we had something truly spectacular on our hands," he said. "The feeling of elation in the OSS area was indescribable, and I doubt that I will ever experience anything like this again. This was not the 'Big Fizzle' that had been predicted only one week earlier, but rather the most dazzling astronomical display of the century."[7]

A few minutes later, Heidi Hammel burst into our news conference with the stunning images from the Hubble Space Telescope. As I watched and listened, I realized that thanks to the spectacular death of this particular comet as it smashed into Jupiter, humanity could never look upon comets in the same way again. Comet Shoemaker-Levy 9 did not go gently into that good night. It ended its life with a tremendous punch into history, one that will strongly encourage scientists, governments, and the public to look at the question of impacts in the solar system with a new level of seriousness. We saw an impact happen on Jupiter; will we see one on Earth? Someday we will.

TWENTY

1999: FINDING PLANETS AROUND OTHER STARS

> *It was quite a day when we made that announcement of 47 Ursae Majoris and 70 Virginis. We rarely knew which TV network station we were being whisked to, from hour to hour. We just smiled into the cameras and described the wonderful planets.*
>
> —Geoffrey Marcy, on his team's
> first discoveries of extrasolar planets[1]

With the dawn of a new millennium, one of the most intriguing astronomical stories is how planets are being discovered not in our own solar system, but in others. Here we will trace the story of how these planets are being detected, and how many worlds there might be in our galaxy.

FROM TOMBAUGH'S PLUTO TO BRAD SMITH'S BETA PICTORIS

When Clyde Tombaugh discovered the ninth major planet in our solar system, Pluto, in 1930, it appeared that the story of new planets would end there, possibly permanently (see chapter 14). It is true that no new planets have been discovered in our solar system. But what about others? In the early 1960s, astronomers investigating a wobbling motion in E. E. Barnard's star suggested a planet ten times the size of Jupiter might be affecting its motion through the sky. Work by astronomer Peter van de Camp kept that dream alive until 1973, when new and more accurate data did not confirm the earlier observations. In that time, however, the idea of detecting such a world seemed out of the question. Planetary science was restricted to the planets in our own solar system.

Planetary science was also attracting some of the best scientific minds. One of those minds belonged to Brad Smith, who, as a member of the Army Map Service in the 1950s, was assigned to geodesist John O'Keefe to work on a geodetic survey in which distances between land masses are measured precisely. At the time, the group was using the Moon as the most convenient object for calculating the trigonometric triangulation. By timing occultations of stars by the Moon from two different sites on two continents, Smith and O'Keefe could determine distances between the two sites with an accuracy of a few yards.

O'Keefe had heard that Clyde Tombaugh was developing a plan, through the Army's Ordnance Research, to search for natural satellites of the Earth. If any were found closer to the Earth than the Moon, they would provide far better triangulation, hence far better distance measurements, than the Moon would. But O'Keefe was wrestling with the vexing problem of interservice communication within the army. O'Keefe assigned Smith to work with Tombaugh. Although Tombaugh did not succeed in finding any natural satellites, he and Smith did become close friends, and after Smith's assignment was done, he retired from the army and joined Tombaugh's program. "Brad came to spy on me," Tombaugh loved to say, "and I kidnapped him!"[2]

As a result of this collaboration, Smith became as interested in planetary science as Tombaugh, so he entered the New Mexico State University as a graduate student and completed the first doctoral dissertation in the history of the astronomy department. He later skillfully led the imaging team for the *Voyager* spacecraft during its historic visits to Jupiter, Saturn, Uranus, and Neptune. Two years after the spacecraft passed Saturn, another spacecraft called *IRAS*, or Infrared Astronomical Satellite, mapped the sky in the infrared, recording sources of heat that could not be observed from the ground or from other means. It recorded a hot area surrounding the young star Beta Pictoris.

Brad Smith followed up that discovery in 1984 by making use of a coronagraph—a device that blocks out the bright light from a star—to record anything in the area immediately surrounding Beta Pictoris. What Smith recorded was beyond anything that he could possibly have hoped for. Without the star's bright light to blot it out, Smith found a large area of delicate nebulosity surrounding the young star. Was it possible that Smith had detected something amazing—the early stages of the formation of a new system of planets? Indeed, if that interpretation was correct, it would be one of the most remarkable discoveries of our time. Another possibility existed: It could be a sphere of comets, much like our own Oort cloud, that circles the sun at the very edge of our solar system. Either possibility, or both, could be correct. In either event, Brad Smith has discovered an awe-inspiring new phenomenon, the possibility that we are witnessing the birth of a solar system.

THE SEARCH BEGINS

Brad Smith discovered the possibility that Beta Pictoris might have a nascent system of planets circling it. He did not actually discover a planet circling another star. In the autumn of 1986, Geoffrey Marcy, a young assistant professor at San Francisco State University, and Paul Butler, a physics graduate student with an undergraduate degree in chemistry, joined forces to begin an actual search for planets that orbit other suns. Because all these planets are about a billion times fainter than the stars around which they orbit, detecting them required a new approach that makes use of an old law. "For every action," Newton's Third Law says, "there is an equal and opposite reaction." As the Earth travels around the Sun, the Sun swings very slightly in the opposite direction.

Because the Earth is so much smaller than the Sun, this wobble is very small. If, however, the wobble of a distant star is measurable, it could point to the existence of a planet: One wobble equals the planet's year. The wobble can be detected when astronomers measure a star's radial velocity, which is a measure of the star's motion toward or away from the Sun. Radial velocity is measured in the same way that Hubble measured the distances to distant galaxies, from the Doppler effect on the star's spectrum (see chapter 13). In order to measure that wobble, Marcy and Butler would superimpose spectral lines of a particular chemical onto the spectral lines of stars as observed through a telescope.

Butler's initial task was to find a substance that could produce useful spectral lines for calibration. One chemical that turned out to be a good candidate was thiophosgene, which happens to be closely related to phosgene, the mustard gas that was used in World War I. Fortunately, the far more benign element iodine turned out to be the best; its spectrum would make an excellent calibration tool. With this insight, Butler built an iodine cell (or container) for use at Lick Observatory, and in 1987 the team's search for planets began.[3]

The observing that Marcy and Butler did was a far cry from standing in the bitter cold on a mountaintop in winter. Their assigned observing period was about three nights per month, sitting at a computer in a heated room some fifty feet from the telescope.[4] Butler spent the rest of his time modeling his data in an effort to get more precision out of it. "We honestly did not know if our tactics would pay off," he says. "We did not know if we would ever achieve sufficient measurement precision to detect planets orbiting other stars."

PLANETS AROUND A DYING SUN

The obvious places to search for planets would be, as Marcy and Butler thought, around stars that are similar in nature to our Sun. This suggests that stars that closely match our sun's spectral type, G2, should be the ones most likely to have planets that we would be interested in. Our Sun's G2 classification means that it is near the middle of the various types of stars. It has some ultraviolet radiation, but not too much, yet it is not as red as a star like Arcturus. Thus it was with some astonishment that astronomers found the first strong evidence of an extrasolar planet not from such a star, but from the burnt out remains of a star that had gone supernova many thousands of years ago. All that is left of this once blazing sun is a small neutron star that pulses as it spins very rapidly. This pulsar goes only by the name of B1257+12, a code representing its position in the sky. In 1991, Alex Wolszczan and Dale Frail found evidence of three planets, each one about the size of the Earth, orbiting about the spinning remains of this star.[5] In the decade since that discovery, searches for planets have centered around stars like our Sun, not around pulsars. Frail has gone on to study, through the Very Large Array radio telescopes in New Mexico and through the Chandra X-ray telescope in space, exotic gamma ray bursters that emit, for an instant, the same energy as an entire galaxy. Frail helped show that a burster is a huge (thirty solar mass) star collapsing into a black hole. Frail's work is with modern, computerized instruments on Earth and in space, not with cold nights and old telescopes. "I marvel at the ability of people like you and Roy Bishop," Frail recently wrote to me, "who observe until late in the night. I prefer a good night's sleep and automated telescopes like Chandra and the VLA!"[6]

Frail's mysterious worlds pose some important questions. Assuming they existed long before their sun went supernova, what were they like? Is it at all feasible that life might have existed there? Even more fanciful, what would it have been like to be on that planet, witnessing the instantaneous collapse and explosion of its life-giving star into a supernova? In any event, the blast would have instantly snuffed out any life forms those planets offered, and these worlds are so far away that we may never answer these questions.

51 PEGASI: A STAR LIKE OUR SUN

The search next made the news in 1995, when Michel Mayor and Didier Queloz, of the Geneva Observatory, detected a planet around the nearby star 51 Pegasi. This star is no pulsar; 51 Peg is a bright G0 star, just a little bit bluer than our own Sun. The planet orbiting it is more than half the mass of Jupiter, yet completes its orbit every four days and five hours! For the planet to circle the star that fast, that world would have to orbit virtually in the outer atmosphere of the star. The closest planet to the Sun in our solar system is Mercury, which takes eighty-eight days to complete one revolution.

In October 1995, Marcy, now at the University of California at Berkeley, and

Butler, now at the Carnegie Institution of Washington, confirmed the existence of a planet around 51 Pegasi. The discovery of 51 Pegasi's planet was a bit of tough luck for the pair. They had spent the previous eight years carefully refining a procedure that would yield very precise results, but since the procedure was very costly in computer time they had not yet reduced the bulk of the data they had already collected. The Geneva team, using somewhat lower precision, got their results much faster. Nonetheless, Marcy and Butler were pleased with how things turned out. "We felt lucky simply to be in the running to detect extrasolar planets . . . that we were even on the right planet-hunting track. We had the wonderful pleasure of quickly confirming the 51 Pegasi result within one week. We were extremely excited that our dubious subfield, planet hunting, was suddenly deemed mainstream science."[7]

THE WORLD CIRCLING 47 URSAE MAJORIS

The race was on. During the late fall of 1995 Marcy and Butler put as much computer power as they could find into the work of analyzing the data their telescopes had already acquired. They didn't have to wait long before their first discovery: a big planet three times the mass of Jupiter orbiting the star 47 Ursae Majoris. There were some beautiful things about the orbit of this new planet. Instead of huddling close to its star like the planet of 51 Pegasi, this world orbits at a comfortable 2.1 astronomical units, or 2.1 times the Earth–Sun distance, from 47 Ursae Majoris—a finding that was not what they expected. "I felt scared," Marcy recalls that time.

> The [star's radial velocity motion] showed an apparent wobble of the star with a 3-year period. I didn't believe it, but Paul did. I couldn't bear the possibility that we might announce a wonderful planet in a nice, 3-year nearly circular orbit, only to be embarrassed later by a retraction if we were wrong.
>
> So, Paul and I took more data that December 1995 and the velocities fell exactly where we predicted, due to the planet that we thought was there. We were thrilled, but then scared for another reason. At the [American Astronomical Society (AAS)] meeting in San Antonio, to occur the next month, I was slated to give an invited talk on extrasolar planets. The invitation had come even before the announcement of 51 Peg. So this was miracle timing for a talk. But we were deathly afraid that Michel Mayor also had 47 UMa on his observing list, and would scoop us, while we waited for the AAS meeting to occur. But we waited for three weeks, from late December through early January. It was quite a day when we made that announcement of 47 Ursae Majoris and 70 Virginis. We rarely knew which TV network station we were being whisked to, from hour to hour. We just smiled into the cameras and described the wonderful planets, only to learn later that *Time* magazine would put the discovery on the cover.[8]

I was personally surprised that a planet might exist in the 47 Ursae Majoris system, some forty-two light years away. The star, a little south of the Big Dipper, is faintly visible to the naked eye. When I was a teenager I was assigned a small area of the sky to search each night with binoculars for possible comets. I have been familiar with 47 Ursae Majoris since I first looked at it on May 1, 1964. Now I learn that a planet at least twice as large as our own Jupiter orbits that star, once every three years. It is unlikely that anyone can live on that planet, but maybe it has an Earth-sized moon that supports life, or there might be an Earth-like planet in the 47 Ursae Majoris system. Now when I look at that star, I wonder if someone there is looking back.

As of the end of the year 2000, the number of extrasolar planets had risen to about fifty, with a few more still as suspects. Marcy and his team are the world's leading extrasolar planet discoverers so far. They did not come in with the first planet, circling 51 Pegasi, but they did discover five of the first six new extrasolar planets. The first eight to be found circled stars like Upsilon Andromedae, Rho Cancri, Tau Bootis A, 16 Cygni B, Rho Coronae Borealis, and 47 Ursae Majoris. Although these planets are much larger than Earth and probably don't have any life, it is certainly possible that all these sunlike stars do have other planets that are close to the size of Earth and that some of these could be homes to life.

Of all their finds, the system of three worlds orbiting Upsilon Andromedae is the most special to Marcy. "Originally, in 1987, we never expected to find even one planet around any star," says Marcy. "To find a system of two planets would have been marvelous. But the first system we ever found had three. For three years after discovering the inner planet around Upsilon Andromedae, Paul Butler and I knew that the velocity wobbles were not consistent with that planet. Something was wrong with that planet. We thought that maybe the inner planet was a mistake on our part. It took three years to figure out that our mistake was in ignoring two more planets!

"The first extrasolar planetary system is the nicest catch I can imagine."[9]

GLIESE 876 RESONATES

–At least for now. In January 2001, Marcy and Butler announced two additional solar systems. Headed by a star called HD 168443 in the constellation of Serpens, the first system is about 123 light years from Earth. One of its two planets is seven times the mass of Jupiter, the other is seventeen times. Could these planets (especially the larger one) really be brown dwarf stars, an intermediate type of object between a large planet and a star that has begun nuclear fusion? Marcy and Bulter say no, since these planets probably both formed in the disk of material surrounding the star during the early history of the system. Why these planets grew so large is a mystery.

Gliese 876, the second system, consists of two planets orbiting a red dwarf star in Aquarius. Next to brown dwarf stars, which do not undergo nuclear fusion, red dwarfs are the lowest temperature stars of which we know. Because they are so faint we know only of the closest representatives; Gliese 876 is only fifteen light years away. One of its two planets has an orbital period of thirty days, and the other, at sixty days, is exactly double. Thus, the planets are in a two-to-one resonance. The system is somewhat similar to the outer portion of our own solar system, where Neptune and Pluto orbit in a three-to-two resonance. Debra Fischer, a member of the Marcy–Butler team, suggested that the two-to-one resonance is like a violin that can play a single note with two harmonics.[10]

PLANETARY TRANSITS

For the first nine years of the extrasolar planet age, from 1991 to 2000, the only real evidence for these new worlds was the wobble they caused in the radial velocity motion of their host stars. Things changed during August 2000. David W. Latham, a prominent scientist at the Harvard-Smithsonian Center for Astrophysics, had analyzed the motion of the star called only HD (for Henry Draper Catalogue) 209458 and concluded that a companion, most likely a planet, orbited that star. Moreover, the orbit was oriented edge-on toward Earth. This meant that once each orbit the planet would pass in front of its star, and during each transit would cause the light from the star to drop suddenly for a short while, then rise again.

On September 9 and again on September 16, 2000, Tim Brown and David Charbonneau recorded exactly the drop in the star's light that Latham predicted. This new method allowed astronomers to learn far more about the nature of the new planet than would be revealed by the other methods. For example, from the transit, the astronomers calculated that the radius of the planet circling HD 209458 is 91,000 kilometers, or 1.27 times the radius of Jupiter. In turn, the planet's density is less than water, so that this world, like Saturn, would float in water (if a body of water large enough could be found).

Observing planets transiting their stars could offer a goldmine of information about these new worlds. However, only those worlds oriented toward Earth will work; if the orientation isn't right, we never see a transit.[11]

SEARCHING FOR LIFE ON THESE WORLDS

The ultimate goal of all these searches for planets is to discover a planet about the size of Earth, orbiting a star like the Sun, a planet capable of supporting life. On closer examination we might find that this Earth-like planet contains continents, oceans, mountains, valleys, and rivers. It would be difficult to believe that a planet with all those features would not also be home to some sort of life —not necessarily intelligent life with technological capabilities, but some life form.

What if we find evidence that there is a civilization out there? What if someone is sending signals? We do, in fact, know of one such planet. It's called Earth, and it has been sending intelligent signals out into the galaxy, at the speed of light, for a century. There are all manner of potential receivers within 100 light years of the Sun. It is possible that the galaxy is teeming with life, and that some of that life knows of us. After all, we've been freely advertising ourselves for a century, since the Italian physicist Guglielmo Marconi sent a wireless message across the Atlantic in 1901. That simple message traveled across an ocean but did not stop at the other side; it went on around the world and into space. By the end of the First World War it had reached sixty-seven star systems, and by now it has passed through many hundreds more.[12]

We know of one successful planet in one solar system; are there others? To answer this question, we should consider that only a small fraction of other stars are the size and type of the Sun. A brilliant blue supergiant star like Rigel, in Orion, emits so much ultraviolet radiation that life as we know it should be impossible. Even our own Sun has UV radiation, but thanks to the development of the protective ozone layer in Earth's atmosphere, life can exist here anyway.

AN EQUATION FOR LIFE

Although science fiction has the galaxy teeming with intelligent species, no extraterrestrial life forms (with the possible exception of a single rock from Mars that found its way into meteorite hunters' hands in 1984), of any kind have been found, let alone intelligent life. Moreover, we have no idea whether we could communicate with other civilizations even if we could detect them. We have no way of knowing, for example, what humpback whales tell each other.

In 1961 the American astronomer Frank Drake considered that problem from a mathematical viewpoint that looked at a series of possibilities. The version I use here was adapted by Carl Sagan, and I have added a new variable called f_j. The equation follows:

$$N = N_* f_p \, n_e \, f_j f_l \, f_i \, f_c \, f_L$$

Of all the values in this equation, we have a fair idea of only one, N_*, which is the number of stars in our galaxy. Estimates range from 100 billion to 400 billion; we use 200 billion. All the rest are variables designed to focus our search.

- The fraction f_p limits our search to those stars that have planets circling them. It is possible that every star has some sort of planetary system. But if we say that only half the stars do, then our search is now limited to 100 billion stars.

- The phrase n_e means "ecologically sound for life." It answers the question of how many planets orbit within a particular distance from a star which would be conducive to life. These planets should not be like Jupiter, which might not even have a solid surface, although Carl Sagan has suggested that air-borne life forms could exist even in an atmosphere of a gas giant like Jupiter or Saturn, or practically any of the new big planets recently discovered in other solar systems. If a tenth of the 100 billion suns with planets had at least one of this type of world, we'd be left with 10 billion life-zone worlds.

- f_j: Gene Shoemaker once suggested to me that it is not enough for a planet to exist in a life zone; at least one large planet like Jupiter needs to exist somewhere in the system as well. In Earth's case, were it not for Jupiter and its strong gravitational force being the solar system's vacuum cleaner (see chapter 2), the Earth would still be a target for major impacts every millennium. No form of advanced life could evolve in such a case. If a tenth of the solar systems comply, we now have 1 billion possibilities. With so many discoveries of Jupiter-sized planets, the fraction might be higher than that.

  There is another side of this big-planet issue. It may be common that these planets, like the ones believed to circle 16 Cygni B and 70 Virginis, are in elliptical orbits that would cause them to occasionally interact, or even collide, with Earth-type worlds trying to orbit in a star's life zone. In fact, were Jupiter not in a nearly circular orbit, Earth and Mars would probably have been destroyed or flung out of the system long ago.[13]

- f_l: the fraction of planets where life does take hold. The value of this fraction is the subject of hot debate: Some scientists think that life will probably arise on all these worlds; others say that the origin of life is a rare thing. Let's split the difference, and suggest that life starts on half of these planets. We are now left with 500 million worlds on which life has begun.

- f_i: The difference between life of any sort and intelligent life is enormous. If we say that intelligent life evolves on only one-one-hundredth of these planets, then f_i brings us down to 5 million intelligent worlds.

- f_c: The ability of a species to communicate is our next consideration. If a tenth of these intelligent worlds develop an ability to communicate with us through technology, then the number drops to 500,000.

- f_L: Although we have the ability to communicate via radio and television, we've had it for only a century. The last part of the equation asks how long a technological civilization will last. In our experience, we developed the ability for technological communication just a few decades before we acquired the ability to destroy ourselves with nuclear arsenals. Is that a coincidence, or is the galaxy full of civilizations that have radio and that are on the brink of self-immolation? The fraction f_L is a way of looking at the sociological state of the galaxy. It would be wonderful if most civilizations do survive after reaching the capacity for self-destruction. Let us say that of all the civilizations we have postulated so far, a tenth of them are alive and well right now. Then we should be able to hear the evening news from 50,000 civilizations.

Actually, 50,000 civilizations capable of communicating with us at the present time seems a very large number. It is far less likely that only one planet in our entire galaxy, Earth, has allowed life as we know it to evolve, but such a scenario is possible.

COULD N = 1?

It is possible that life on Earth is the result of a unique set of circumstances, some of which are completely accidental, that would be extremely unlikely to reproduce elsewhere? The following items might be unique to Earth:

- The zone around a star in which a planet might support life might be extremely narrow. If the Earth were just a million miles closer or farther from the Sun, life as we know it could be far more rare.
- The Earth has a very large Moon, which helps to stabilize the tilt of our planet. If we had no Moon, the Earth's tilt might vary, an effect that would lead to disastrous seasonal variations in our climate. As it is, the Earth's tilt is stabilized at a not-too-ideal 23.5 degrees. According to astronomer Clyde Tombaugh, were the tilt less, seasonal variations with their resulting severe storms like tornados in spring and hurricanes in fall would be less severe.[14]
- Earth has a smaller planet, farther out from the Sun, on which life might have had an easier time getting started. Mars is that world. Once life began on Earth, it would be subject to frequent extinctions with each successive comet impact. Since Mars is a smaller target, it would have been less subject to bombardment than Earth, and if life did get seeded there, it might have even been transplanted here through other impacts. This idea of panspermia, formerly ridiculed, is actually gaining some acceptance in the community of planetary scientists. If this is so, microscopic life forms could have traveled as passengers on rocks hurled from Mars, the results of impacts there, and headed to Earth.

- The amount of carbon available on a world is a critical question. There needs to be enough for life, but not so much that a runaway greenhouse effect takes place, as Venus's temperatures of higher than 800° F have shown.
- Somehow, photosynthesis became established on Earth, at just the right time and in the right amount, and subsequently played a major role in the evolution of life. Is that common, or virtually unseen in our galaxy?
- For some reason at the beginning of the Cambrian period some 543 million years ago, there was a veritable explosion of life on Earth. Almost all of the animal phyla made their initial appearance in a relatively short period starting at that time. Is this unique to Earth, and does the evolution of intelligent life depend on whatever took place then?

For these and other reasons, planetary scientists Donald Brownlee and Peter Ward suspect that complex life, particularly intelligent life, might be rare in the universe. If that is the case, then life as we have it here might be even more special than we thought.[15]

SEARCHING FOR EXTRATERRESTRIAL INTELLIGENCE

A worldwide effort to search for extraterrestrial civilizations has been underway for some years. Called SETI, for Search for Extraterrestrial Intelligence, the program uses a variety of radio telescopes to search for artificial signals from other worlds.[16] Although some unexplained signals have appeared, none have yet confirmed the presence of life elsewhere. SETI programs are being conducted at several institutions, including Harvard and Berkeley. One program is unique. Called SETI@home, the program is a search for extraterrestrial signals using the National Astronomy and Ionospheric Center's 305-meter-wide Arecibo Radio Telescope. The survey covers almost 30 percent of the sky at 1420 megahertz, which is the 21cm hydrogen line discussed in chapter 16. Most researchers believe that if extraterrestrial civilizations are sending signals, then the 21cm hydrogen line would be a natural frequency for them to use.

Organized by the University of California at Berkeley's SETI program, SETI@home utilizes hundreds of thousands of home computers all over the world. (For more information, visit http://setiathome.ssl.berkeley.edu/. Instead of running an ordinary screensaver program, these computers run a screensaver that actually uses the computer to crunch the vast amount of data obtained by the telescope. By the end of 2000, half a million years of computer time joined the search for extraterrestrial intelligence as part of this program.

The magic of SETI@home lies in its use of so many ordinary home computers that otherwise would be turned off or lazily running screensaver programs. Each of the owners of these computers is left with the feeling that he or she is accomplishing some basic scientific research, since the Berkeley SETI program now has the computer power to analyze the enormous quantities of data it has gathered. If the program is successful someday in detecting and confirming the extraterrestrial signal sent from an intelligent civilization, it is possible that your computer, if you join, might have been the one that analyzed the crucial data.[17] If the discovery is ever made, everyone participating in the SETI@home program would share that exhilarating feeling of participating in a discovery.

A few years ago, nine known major planets existed in a single solar system. With the number of known worlds increasing rapidly, and with searches for extraterrestrial intelligence expanding, it is not out of the question that one day we will detect the signals of some remote intelligence, and we will also get a picture of the distant world that that civilization calls home.

TWENTY-ONE

1971: ACQUAINTED WITH THE NIGHT

Summerfield and O'Meara

One luminary clock against the sky . . .
I have been one acquainted with the night.

—Robert Frost, 1928[1]

On the night of October 27, 1971, I watched the Moon from an old dyke near Nova Scotia's Minas Basin. The long earthen structure was similar to one that the Acadians had built two centuries earlier before their expulsion from the area, and I thought of how they might have enjoyed a dark night sky from this maritime location. Both the Moon and I were surrounded that night; the Moon by a halo, and me by silence, darkness, and space. As I turned on my tape recorder, I started to hear the soft piano sounds of Beethoven's Moonlight Sonata. Never was I so transfixed by a confluence of sight, sound, and the feeling of poetry that only the wondrous darkness of the night sky could write.

Of the many stargazers who share my spiritual feeling for the sky, I profile two of them in this chapter, Robert Summerfield and Stephen James O'Meara. Summerfield loves the night sky for one reason, to share it with children and instill within them the feeling for discovery that he has. Summerfield knows that when a child first looks at the Moon, Jupiter, or the Veil Nebula through one of his big telescopes, the feeling of discovery is as strong as though the child had been the first to discover a new world. In a sense, the child had indeed discovered it. O'Meara approaches the sky differently; his eclectic interests span the natural world. He chases volcanoes to watch their eruptions; he acquaints himself with the lifestyles of alligators and bats, and he watches the stars.

187

When you tour the stars the way these people do, you let the stars come to you. Some of the best things I've ever seen in the night sky have come not from nights at the telescope but during quiet sessions sitting on the Minas Basin dyke, or walking under the stars. The brightest meteors I've ever seen have come my way during such starlight strolls using neither binoculars nor telescope. Reverend Thomas Anderson was also one such walker. One February night in 1901, his eyes watching the sky, he found a star that didn't belong among its neighbors in the constellation of Perseus. Now known as Nova Persei 1901, this star is still visible through a telescope, changing slightly in brightness over time.

Some of us, like Summerfield and O'Meara, are attracted to astronomy not so much for its subject matter but for its environment. "The kingly brilliance of Sirius pierced the eye with a steely glitter," Thomas Hardy wrote in one of his greatest novels, *Far from the Madding Crowd*, "the star called Capella was yellow, Aldebaran and Betelgeuse shone with a fiery red. To persons standing alone on a hill during a clear midnight such as this, the roll of the world eastward is almost a palpable movement."[2]

ASTRONOMY TO GO: DISCOVERY THROUGH TEACHING

A great attraction of astronomy is the room it allows for the imagination to roam. No matter how we choose to excite the next generation about the sky, it could be the sight of a meteor in the night, or a meteorite that Bob Summerfield brings into a classroom, that actually does the trick. It could also be a sense of place that a look at the Moon might bring. Thomas Hardy lived in a time, a century ago, when the Moon was a place only for poets. Robert Frost died in 1963 when going to the Moon was still a dream. But in today's world, the Moon can be seen as a physical place, for we have seen the images of a dozen men striding across its rocky surface.

People like Thomas Hardy, Robert Frost, and Bob Summerfield don't do the kinds of observations that result in the discoveries of quasars or comets. But the discovering they have done is no less important. Through their prose, poetry, and teaching, they have elicited the siren call of discovery in the rest of us. The meteorite Summerfield brings into classrooms is part of his program called Dead Aliens. Not the ETs the kids were expecting, these particular "aliens" are the meteorites, the aliens, indeed, who have fallen to Earth to stay. The children are captivated as much by Summerfield's energy and enthusiasm as they are by the fascinating world of meteorites.

Born in Philadelphia on February 8, 1961, Summerfield's first astronomical memory was the eclipse of the Sun on March 7, 1970. His third grade teacher,

Barbara McAdams, piqued his interest by writing a large number on the board, with many, many zeros. Summerfield has long forgotten what the number was about, but it was enough to instill in his memory that the universe is a big place. When the teacher gave a costume party for which each child had to dress as a constellation, he dressed up as Auriga the charioteer.

By his senior year, Summerfield was cutting classes to help his teacher friend Scott Negley in the high school's planetarium. By this time Negley and Summerfield had become close friends, and Summerfield was developing a passion for telescopes. In 1986 he attended the Stellafane telescope maker's meeting on its famous hilltop near Springfield, Vermont. John Dobson was the keynote speaker that year. Dobson invented a popular, easy to construct kind of Newtonian reflector that made it possible for almost anyone to buy or make a large telescope for a reasonable amount of money. For Summerfield, Dobson at Stellafane equaled heaven. "Russell Porter, who founded Stellafane in 1925, made astronomy available to the amateurs," says Summerfield; "Dobson made it available to everyone else."

In the spring of 1990, after looking through a twenty-five-inch reflector telescope that Dobson inspired, Summerfield started Astronomy to Go. "With a public telescope this big," he reasoned, "more people would come and look. Just imagine how the public would react to a gigantic telescope in a parking lot!" Since then thousands of people have looked through Summerfield's collection of telescopes, which now is topped by a thirty-six-inch-diameter behemoth called the Yardscope.

In the mid-1990s Summerfield met Lisa Levikoff, a student who began to assist with his program. The couple married in 1995, and since then have taken several cross-country trips in their telescope-filled van each year. On one of their trips their overworked vehicle simply gave up, caught fire, and burned to a crisp. The couple was unhurt, but one of their telescopes was destroyed and several others were damaged. As they stood by the roadside watching their life's work burn away, they never thought for a minute that the children in one of the classes they had visited would raise more than $200 to replace the ruined telescope. The meteorites survived, Summerfield noted ruefully, "because they've been through fire before."

Four days after the fire, using a borrowed telescope, the Summerfields kept their appointment with a school that had asked them to bring telescopes for a star party. Using the borrowed telescope they showed the children the stars. "We have never canceled a single program," Summerfield says with pride.

What if the sky is beautifully clear, it's new Moon weekend, and he wants to observe faint objects by himself? It doesn't happen. Summerfield *never* wants to observe on his own. "Certainly not if a group calls," Summerfield answers passionately. "Sharing the sky with the public is what my whole life is all about."

PASSION IN NATURE

When Summerfield shares the sky with his friend Steve O'Meara, sparks really do fly; together the two prominent amateur astronomers represent the best of "amateur" astronomy. After all, *amateur* comes from the Latin word for love. For O'Meara, discovery in astronomy need not be of a new object. To learn something new about an already known object is at least as important, and O'Meara specializes in such findings.

I met O'Meara for the first time in Cambridge, Massachusetts, at the 1978 annual meeting of the American Association of Variable Star Observers, an organization to which hundreds of observers contribute their observations of the behavior of variable stars (see chapter 11). We had been sitting through an afternoon of short papers, and as the sky darkened at the end of the day, O'Meara was to deliver the final paper. A young college student dressed in black, Steve O'Meara had ten minutes to summarize the life and times of George Bond, one of Harvard College Observatory's first directors. O'Meara gave such a passionate story of the man's tenure that for those minutes we, in the audience, felt as if we had been whisked back to a time long gone.

I soon learned why O'Meara's presentation was so impassioned. In a time when observing practices were changing rapidly, O'Meara wanted to hang on to the old visual methods that triumphed in the past. "I'm a nineteenth-century observer who happens to live in the twentieth century," O'Meara describes himself. He especially likes the traditions of people like the comet discoverer E. E. Barnard, who spent most of his night hours at the eyepiece of a telescope. These people lived in simpler times, when science was done not with large teams in warm control rooms, but on lonely mountaintops and dark farmers' fields.

O'Meara's view of the stars stems from a larger interest that he has nurtured since childhood, to observe nature's various manifestations. In seventh grade he built a miniature volcano from plaster of Paris, chicken wire, and sulfur. During high school O'Meara's interest gravitated so strongly to astronomy that he somehow wrangled permission to use Harvard's beautiful nine-inch refractor. Built by the nineteenth-century firm of Alvan Clark and Sons, this telescope has first class optics, plus enough light gathering power to provide fine images of the planets. In preparing himself to observe properly, O'Meara trained his eyes to the point where he could see very faint stars in addition to details on planets well beyond the limit of visibility for most people.

Thus, the legend grew of "CCD-eyes O'Meara." Over the years O'Meara drew several sketches of Saturn in which he detected thin lines stretching across the rings like spokes in a bicycle wheel. Although his detection was ridiculed at the time, O'Meara pointed out that other observers had also detected spokes in Saturn's rings. In November 1980, *Voyager 1* swung past Saturn and photographed the rings in far greater detail than had ever been seen before. Instead of three rings, there were hundreds—and through them all were O'Meara's spokes. Not only did *Voyager* confirm the spokes, but the spacecraft also imaged them staying together as the ring particles orbited the planet, a phenomenon the Voyager Imaging Team called the "Saturn 500."[3]

FINDING HIS PASSION IN A DISTANT LAND

Indulging his seventh-grade passion for another part of nature, in September 1982, while observing at the 14,000-foot-summit of Mauna Kea on Hawaii's Big Island, a telephone call brought news that Kiluaea volcano, on the same island, was erupting. O'Meara could even see a distant red glow against the clouds. Dashing from one science passion to another, two hours later he was hiking into the caldera (the basin located inside the cone of a volcano), and got close to lava that was not orange but a blindingly white liquid rock as bright as the Sun. "Suddenly I heard soft tinkling noises like the breaking of fine crystal. That's when I looked down at my feet, and realized that I was a foot away from an advancing river of lava!" Backing away to relative safety, O'Meara watched nature's awesome show in silence. He lifted a boulder the size of a volleyball and threw it in the molten river, expecting the lava to plop around it like water. "The lava buckled a bit," he reported, "as if it were made of rawhide, then lifted the boulder up and carried it away. The top of the flow was already hardening."

All through the night the Earth spoke, its flaming curtain swaying and swishing like the hula dancers that the islands are famous for. Then toward dawn, the curtain weakened, its graceful dance transforming into a giant piston as the gas content of the eruption increased. The eruption's sound changed from gentle hissing to something akin to a man trying to breathe with a punctured lung. There was a gasp and a choking sound as big boulders shot into the sky, then they rained back down to cover the vent. With a final agonized gasp, the eruption was over. On that night O'Meara listened intently as the Earth opened and spoke to him with a gasping voice.

Two years later, in January 1985, O'Meara was back at Hawaii's Mauna Kea Observatory for a different reason. Instead of searching out a volcano on Earth, he was looking for one in space, a body called Halley's Comet, that was erupting dust and gas. Although the comet had been detected electronically through large telescopes, no one had actually seen it since Max Wolf's last observation in 1911 from Heidelberg. When the comet was near Earth, volcanolike eruptions of dust and gas from its nucleus caused it to brighten. On this night, O'Meara hoped that after three-quarters of a century, the comet was bright enough to be seen by human eyes once again. Using a twenty-four-inch reflector telescope, he detected stars as faint as magnitude 20.4, many times fainter than most observers even try for. He did catch the comet that night at magnitude 19.6, making the first visual observation seven months before anyone followed.

The Halley sighting pales beside O'Meara's series of drawings of Uranus, a planet that offers a blank face to most of us. Several years before his Halley observation O'Meara, in 1981, noticed two specific spots on Uranus, one near the pole, the other closer to the equator. Using Harvard's nine-inch telescope, he drew the planet several times as the features moved across the field of view. Because Uranus is tilted so strangely, rolling through the sky on its side, only one hemisphere is pointed toward us. On a normal planet like Jupiter, features appear to move straight across the planet's disk, but on Uranus, O'Meara knew that their motion would appear very different. The spot near the pole appeared to move hardly at all, but in a narrow semicircle. The other spot moved in an arc as the planet rotated. "I knew Uranus's axial tilt when I was making the observations," O'Meara writes. Another observer also tried to monitor the motion of the spots, but "he did not think about Uranus's axial tilt and assumed it was similar to that of Jupiter."[4] Once O'Meara took into account the tilt, the spot motion seemed to satisfy a period of about 16.3 hours. In 1984, as *Voyager 2* sped toward Uranus, O'Meara was strongly advised to publish his result, and it appeared on an International Astronomical Union Circular that February.[5] Early in 1986, *Voyager* confirmed O'Meara's period to within ten minutes! Some of the more staid astronomers back then did not believe his earlier observations, but with Uranus there could be no doubt that O'Meara's vision was accurate.

In the spring of 1996, several observers including O'Meara, Jim Scotti, and I observed Comet Hyakutake's marvellously long and beautiful tail. We all estimated its length as 110 degrees, indicating that the tail stretched more than halfway around the visible sky. These tail-length estimates were denounced as absurd by some theoreticians whose models of the comet indicated that the tail couldn't possibly be that long. There the matter rested until the end of 1999, when data from the *Ulysses* spacecraft indicated that the comet's tail might have been longer than any other in observational history, and that it was at least as long as we had said it was.

As O'Meara became better known, partly through his observations and partly due to his job as *Sky & Telescope*'s editor covering amateur astronomers, he made frequent appearances at star parties, sharing his enthusiasm with other observers. It was at the Texas Star Party of 1994, held near the west Texas village of Fort Davis, that the Sun would be in annular eclipse from nearby Carlsbad Caverns in neighboring New Mexico. O'Meara wanted to watch the eclipse from the caves in order to catch the flight of the bats, an event normally reserved for twilight. O'Meara hoped the flight would also happen with the darkening of the sky during the eclipse. Clouds over the cave forced him to cancel

Edwin Hubble at the forty-eight-inch Schmidt camera atop Palomar Observatory.
(Courtesy Mt. Wilson Observatory)

Carolyn Shoemaker at her stereomicroscope, the instrument with which she discovered 32 comets. *(Photo by Jean Mueller)*

Albert Einstein at work. *(Courtesy Mt. Wilson Observatory)*

Steve O'Meara peers through a telescope equipped with a special filter for safe viewing of the Sun. *(Photo by Wendee Wallach-Levy)*

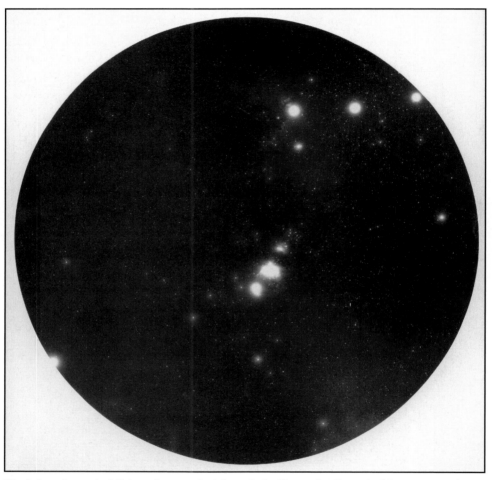

The belt and sword of Orion, photographed through the Shoemaker-Levy double cometograph. *(Photo by David H. Levy and Keith Schreiber)*

Arthur C. Clarke and Tyrannosaurus Rex friend. *(Photo by Rohan de Silva)*

Bart Bok and roadrunner friend relax on the Priscilla Bok Memorial Bench at the Arizona-Sonora Desert Museum. Bok often visited the bench when he wanted to think about Priscilla. *(Photo courtesy of Bart Bok)*

Jean Mueller about to inspect a photographic plate she took for the second Palomar Observatory Sky Survey. *(Photo by Carolyn Shoemaker)*

One of Jim Scotti's early CCD images of Shoe-maker-Levy 9, taken March 30, 1993. *(Photo by Jim Scotti)*

The discovery photographs of Comet Shoemaker-Levy 9, taken 1h 47m apart on the evening of March 23, 1993. *(Photo by David H. Levy, Gene and Carolyn Shoemaker)*

While living in Slough, Herschel often visited King George at Windsor castle. *(Photo by Wendee Wallach-Levy)*

To test the optical quality of his telescopes, William Herschel focused them onto this wall of Windsor castle. By alternately blocking parts of the mirrors and seeing how clearly the stones appeared, he could easily identify problem areas in his mirrors and improve them. While the Windsor Castle method is no longer used, it seemed to work very well for Herschel. *(Photo by Wendee Wallach-Levy)*

those plans, and he sped westward along the eclipse path, looking for a break in the clouds. Running perilously low on fuel, with minutes to go before the annular eclipse, he finally found a roadside spot. Just as O'Meara was setting up his camera, a police officer ordered O'Meara and his group to move their car or face arrest. Somehow they moved their car, jumped out in the nick of time, and caught the ring of sunlight perfectly shaped around the Moon.

That evening at the Texas Star Party, Bob Summerfield stayed up late as hordes of observers looked at distant galaxies through his mighty Yardscope. O'Meara also looked through the many telescopes on the field. For Summerfield, helping the observers at the Texas Star Party peer ever deeper into space is discovery enough. For O'Meara, being a part of nature, whether confronting a volcano in eruption or being part of the celestial lineup of Earth, Moon, and Sun that defines an eclipse is what discovery is all about. It takes the enthusiasm and drive of these men to be truly acquainted with the night.

TWENTY-TWO

1988:
JEAN MUELLER

Of Supernovae and Songs

It was a librarian's dream. I was to help produce one of the fundamental reference tools of astronomy for the next half century.
—Jean Mueller, starting work with the
second Palomar Observatory Sky Survey

Palomar Observatory is a wonderful place to call home. A small California community of maintenance staff, administrative people, and telescope operators live there, enjoying the peace of the mountain. Scattered across the mountaintop are a few small houses that have been used by staff members since the observatory's construction in the late 1930s.

Jean Mueller, a former librarian turned telescope operator, lives in one of the smaller houses. Her lifestyle is a simple one; her twelve-hour shifts are followed by a daytime sleep that gives her precious little time to prepare for the next night, especially during the long nights of winter. Aside from bimonthly drives to the city of Escondido, forty-five-minutes away, and occasional trips to see her family or friends or to chase an eclipse, Mueller confines her life to the woods, and the stars, of Palomar.

Mueller is not an astronomer in that she has not had any academic training in astronomy. Her accomplishments in astronomy, however, are formidable. Through the work she has done at Palomar, she has discovered no fewer than 110 supernovae, just 11 short of the record for a single observer, still held by Fritz Zwicky, the supernova specialist who observed at Palomar's 18-inch Schmidt camera 50 years earlier. Moreover, in the course of her observations, she has discovered fifteen comets and seven asteroids that someday could endanger the Earth.

SUPERNOVA!

When most stars, like our Sun, get older, they expand and become red giants and then shrink to become white dwarfs. But for more massive stars, the end is utterly savage, resulting in the star's total destruction within a few minutes.

In the 1930s two astronomers working at the great observatories in California, Fritz Zwicky and Walter Baade, developed the basic ideas for exploding suns that are accepted to this day. They were not studying the ordinary novae, where a star suffers an explosion that causes it to brighten by a hundred times, but the supercharged blasts that cause a star to shine, for a few days or weeks, with the intensity of hundreds of billions of suns. In rare cases, it is possible for a star like our Sun to become a supernova. If such a star becomes a white dwarf that is a member of a double star system, it may try to capture hydrogen from its neighbor. Normally, the consequence of this behavior can be a periodic nuclear explosion that causes the star to become an ordinary dwarf nova every few months, or a regular nova every several centuries.

But what if the captured material never ignites? If the dwarf keeps on gathering more and more matter, how massive can it become and still stay together? Some decades ago, the Indian-born astrophysicist Subrahmanyan Chandrasekhar proposed that there is a critical mass beyond which a white dwarf cannot stay a white dwarf. As soon as that limit is reached, the star blows up. If the conditions are right, a white dwarf can end its life spectacularly after living its normal life as a star for billions of years. Zwicky and Baade called these explosions Type I supernovae.

A different variety of stellar explosion is the Type II, which is the end result of a massive star that has lived too hard and too fast. Burning itself out in just a few million years, compared to the 10-billion-year-lifespan of a star like the Sun, this star ends its life with great violence. In one weaker subset, the star fuses its hydrogen, and then its helium, until the core is left with carbon. If all the carbon in the core ignites at once, the detonation may be strong enough to blow the core apart in a supernova explosion.

A more massive star will survive carbon detonation, but in doing so it has bought only a few hundred years of time. Stars more than nine times the mass of the Sun are so hot that their carbon ignites gradually, so it is safely fused to oxygen over several hundred years. Then, in a period of about half a year, the oxygen fuses to silicon. After that process is completed, silicon fuses to form a core of iron. That part of the process takes less than a day.[1]

If this massive star only understood nuclear physics, it would know that an iron nucleus is so stable that it cannot release energy in a fusion reaction when enormous amounts of heat are applied. Instead, iron will absorb energy. But the star tries to ignite its iron core anyway by contracting and heating it. Unable to keep up, the core suffers a final and fatal collapse. In less than two seconds the core crashes in on itself, carrying large amounts of still unused fuel; as the electrons crash into the nuclei of their atoms, they form neutrons and neutrinos, and a new kind of star called a neutron star. Around the collapsed stellar core a fierce shock wave pushes the star's outer layers away at great speed. In the titanic explosion, the star outshines the combined light of all the stars in its galaxy.

The hundred supernovae that Jean Mueller has discovered exploded through one of these processes. With each of her discoveries, astrophysicists are able to increase their understanding of how, in essence, these stars die.

LIBRARIAN TURNED ASTRONOMER

The daughter of a bank branch manager, Jean Mueller started on her road to Palomar around 1973 when she was appointed librarian at the University of Southern California's Gerontology Center. Although Mueller found the study of aging quite interesting, she did not see herself doing that for the rest of her life. Typical of many resourceful young college students, Mueller's college career at Cal State, Northridge, left her with a variety of career choices, each pulling in a different direction. She loved politics, and even began an abortive stint in law school. She also loved to sing, a talent she inherited from her mother and which she indulges over the years. Her original astronomical songs, accompanying herself on the guitar, are a pleasure to listen to. But Beethoven's Choral Symphony, however beautiful it might have sounded in the Long Beach Symphony Chorus, was not going to give her a career.

It was not until 1980, when she took an astronomy class at California's Rio Hondo College, that Mueller really caught the observing bug. With a small planisphere, or star wheel, she set out to teach herself a new star or a new constellation each clear night. As the seasons went by that first year, Mueller steadily became more familiar with the sky. Four seasons later an astronomer friend at Mt. Wilson Observatory casually told Mueller that there was an opening for an observer on the observatory's sixty-inch telescope, one of the most venerable instruments in all of astronomy. "The offer was so far out of left field that I didn't know what to think," Mueller recalls, "but boy did my wheels start turning!"[2] Thus, in January 1983, Mueller joined the Mt. Wilson Observatory HK study of singly ionized H and K calcium lines to monitor sunspot cycles on nearby stars that resemble the Sun and some giant stars. Just as our Sun has sunspots that increase and decrease in number over an eleven-year cycle, these other stars should also.

CELEBRATING A TELESCOPE

As Mueller approached her second anniversary at Mt. Wilson, she learned of the plan to shut down the observatory's most famous telescope, the 100-inch Hooker reflector. In chapter 13 we read of that telescope's audacious opening night, and how in the years that followed, Edwin Hubble had used it to develop his theory of the expanding universe. After the telescope's triumphant sixty-two-year career, the night of June 25/26, 1985, was chosen as its final night of observing. Geoffrey Marcy (see chapter 20) and Victoria Lindsay were the observers that night, and they donated the first half of that night to host a party of thirty-five people to celebrate the telescope. During the night, Mueller read from the first night's telescope log, and then astronomer Tony Misch read from the Prologue from Alfred Noyes's poem "Watchers of the Sky," written six decades earlier, after the telescope's opening night.

> "Tomorrow night,"—so wrote their chief—"we try
> Our great new telescope, the hundred-inch.
> Your Milton's optic tube has grown in power
> Since Galileo, famous, blind, and old,
> Talked with him, in that prison, of the sky.
> We creep to power by inches. Europe trusts
> Her "giant forty" still. Even to-night
> Our own old sixty has its work to do;
> And now our hundred-inch . . . I hardly dare
> To think what this new muzzle of ours may find."[3]

These lines about Milton meeting Galileo hark back to chapter 1, where we described the visit that took place between them in 1638. "We all thought," Mueller notes, "and certainly we hoped, that the telescope would be open again within a year." For reasons of funding, the 100-inch was closed for several years, and then, happily, it was reopened. Mueller had a strong feeling about the reference to "our own old sixty" having work to do. But Mueller would not continue the great work of the sixty-inch. Leaving the HK work to others, Mueller headed south to Palomar Observatory, home of the largest telescope in the United States. Beginning July 8, 1985, just as Halley's Comet was brightening, Mueller's new observing challenge was to be an observer on the second Palomar Observatory Sky Survey, a record of the entire northern sky that would replace and augment the one taken after the observatory opened almost forty years earlier. "It was a librarian's dream," Mueller remembers her excitement at the prospect. "I was to help produce one of the fundamental reference tools of astronomy for the next half century."[4]

Mueller approached her new job with the pride and care of someone who knew she was assigned an eminent task. As she was also hired to do occasional assisting on other telescopes for observers, during her job interview she was asked how she would respond to an observer who, in the event of fog or high winds, wanted to keep on observing just to gather that last precious bit of data. They were delighted when she answered that above all, her first responsibility is the protection of the telescope and equipment. The telescope she was entrusted with was a special one. One of the largest Schmidt cameras in the world, the Oschin forty-eight-inch could collect an impressive amount of starlight, over a wide field of sky, on a single photographic plate.

Once at Palomar, Mueller met Alain Maury, a photographic scientist and first rate asteroid and comet finder who taught her the basics of photographing the sky on the fourteen-inch-square plates. Mueller learned to take long-exposure photographs in three colors, red, blue, and near infrared. Alain Maury hoped that Mueller would follow his lead. Taking the photographs was the job description, but scanning each plate as soon as it was dry was an important extra task. This way, Mueller had the chance to discover supernovae, comets, or asteroids moving fast or in unusual directions. At first, Mueller was so consumed with mastering the basics of handling the forty-eight-inch and doing the manual guiding of the telescope that she had little time to devote to scanning. By 1987, however, the telescope was guiding its own exposures automatically, leaving Mueller time to began her search. She met with her first success that August, with two Apollo-type asteroids, objects that cross the orbit of the Earth. (When the orbit of the second of these asteroids was sufficiently understood that she was permitted to suggest a name for it, she named it Ubasti after the Egyptian cat-goddess, and after her own cat.) In October she discovered her first comet. When she found the fuzzy trail of light on her plate, she excitedly telephoned Gene and Carolyn Shoemaker, who were conducting their own search at the nearby eighteen-inch telescope. The Shoemakers promptly confirmed her discovery, and coincidentally found their own new comet on the same set of films!

Early in 1988, Mueller discovered her first supernova, a star roughly equivalent in brightness to its entire host galaxy and positioned in one of the galaxy's spiral arms. She confirmed the find by comparing the galaxy image to one on the original sky survey films, which showed the galaxy without the supernova. She then persuaded the observers on the 200-inch that night to turn their telescope to the new star and obtain its spectrum, which, showing the characteristic lines of a supernova, formally confirmed her discovery. By the fall of 1988, when Maury returned to his native France, Mueller picked up the torch of supernova searching at Palomar, just as Fritz Zwicky had done a half century earlier.

Mueller lives in an ideal spot for this type of work, in a small house on the observatory grounds. In her unusual backyard is the Oschin Schmidt camera, and when she discovers a supernova with it, the observers at the 200-inch telescope, also in her backyard, are usually happy to confirm it by obtaining a spectrum of the star.

Of the comets she has found, Comet Mueller 1991h1 and 1993a were by far her brightest. Discovered on December 18, 1991, on a single sky survey plate taken five nights earlier with the forty-eight-inch Oschin Schmidt, the 1991 comet had to wait ignominiously almost two weeks before being confirmed since bad weather and the Christmas season gave Northern Hemisphere observers a case of comet apathy. The confirming images were finally taken at the start of 1992. They showed the comet heading for a close rendezvous with the Sun near the end of March.

Mueller enjoys her lifestyle on Palomar, even though it has a lonesome aspect that keeps her from the hustle of urban life that she was so used to in her library years. During her early days at Palomar she even kept a horse in a corral behind her house. By the year 2000, the survey was all but completed and Mueller was transferred to conduct observations with Palomar's newest telescope, an interferometer spread out over the grounds near the front door of the 200-inch (see chapter 17). That position was temporary. In the fall of 2000, Mueller was awarded the position of night assistant for the observatory's largest and most famous telescope, the 200-inch.

The thrill of observing with these monster telescopes has never left Mueller; she feels each night the historic sense of using telescopes that astronomers of the likes of George Hale and Edwin Hubble built and with which they uncovered the secret of the origin of the universe. At the 200-inch, she always works with the observer assigned to the telescope for that particular night. She misses those lonely nights at the forty-eight-inch, when, during the course of an hour-long exposure probing ever deeper into the heavens, she would hope that the photograph would reveal a new supernova or a comet. If the dome was pointed eastward, away from all the other telescopes, she might even start to sing Fauré's *Requiem*. With the opened shutter of the dome revealing a blanket of distant suns, and the metal hemisphere of the dome itself providing excellent acoustics, Mueller performed a perfect celestial merger of music, darkness, and the distant explosion of a dying sun.

TWENTY-THREE

2001: TO DISCOVER AN *IDEA*

The Cosmic Vision of Arthur C. Clarke

And it all began today.
—Jules Bergman, ABC News,
on the launch of *Telstar*,
the first active communication satellite.[1]

O ur journey through the lives of those who have made astronomical discoveries closes with the man who is not an astronomer but who has a telescope, and whose discovery was not of an object but of an idea. For me, that idea achieved reality during the summer of 1960. On the clear nights of that lazy summer I observed the orbital flight of a huge balloon named *Echo*. It was the world's first communications satellite or comsat—a ten-story-tall balloon designed to bounce radio signals passively back to Earth. I was so impressed with this effort that when I received my first telescope that summer, I named it Echo. Two years later, in the summer of 1962, I was using Echo (the telescope) to make predawn drawings of distant star clusters like the Pleiades. That was also the summer that I observed 112 Perseid meteors on a single night. It was the summer of *Telstar*, the first communications satellite with active electronics designed to receive messages and relay them back to Earth.

Telstar represented a great leap forward in comsat technology, since it had equipment aboard to relay actively radio and television signals. Once safely in orbit and tested, it broadcast Julia Ward Howe's "The Battle Hymn of the Republic" back to Earth as it signaled the promise of a worldwide communications link using satellites. *Telstar* even prompted a popular hit song of that name

200

in 1962 that still plays on oldies stations, but as the satellite swung around Earth, *Telstar*'s communications had to be short, limited to about twenty minutes.

Telstar was the first of the hundreds of communications satellites, or comsats, that have since been launched. In 1963, *Syncom I* soared higher and higher until finally it parked itself in an orbit 22,300 miles above Earth, a geosynchronous orbit. In an orbit that high, a satellite's angular velocity is the same as that of the Earth's rotation, so the satellite remains over a single spot on Earth. Although *Syncom I* malfunctioned, it did succeed in starting the geosynchronous satellite revolution. With satellites deployed like that, communications would be unlimited, allowing the people of Earth to come closer to being a global community. Actually, *Syncom I* was in a geosynchronous orbit but not geostationary; its orbit did not lie exactly on the plane of Earth's equator. As viewed from below, the satellite did a figure-eight loop over the course of a day.

Communication today would stop in its tracks without these satellites. Telephone, television, internet, weather forecasting, stock market, military communications—virtually everything uses these satellites. When just one of them goes offline due to a geomagnetic storm or other malfunction, the effects are dramatic.

WATCHING COMSATS THROUGH TELESCOPES

The story of how these satellites came to be, from my personal view, is very much a part of the story of astronomy. Twice while comet hunting or looking at a variable star, I have seen an unexpected star in the field of view. I looked for my sketch pad, leaving the telescope for a moment. When I returned, the field of stars had moved to the west. But the new "star" was still smack in the center of my telescope. It didn't move. Then I understood—I was looking at a distant satellite hanging over me 22,300 miles away. What is it sending back, I wonder—a weather map, a telephone call, a TV show, or some military secret? On a third occasion, I was atop Kitt Peak near Tucson. This time our thirty-six-inch telescope was preparing to record data on an asteroid as part of an attempt to understand its shape; to accomplish this we were recording its change of light as it rotated, and repeating our measurements many times as the asteroid orbits the Sun.[2] On this particular night, just as the telescope homed in on the asteroid we wanted to observe, we noticed a "star" that was moving slowly through the field. I quickly turned off the telescope's drive so that it would no longer keep pace with the Earth's rotation and follow the stars. As the field of stars began to move away, the "star" remained centered in the telescope. We had found another comsat in geosynchronous—and also geostationary—orbit.

THE DREAM OF ARTHUR C. CLARKE

Those orbiting specks of light are the work of Arthur C. Clarke, a visionary who in this real way has brought space to Earth. It was he who, decades earlier, first came up with the idea of using space to provide the people of Earth with a global communication. Even though I was brought up on his novels—my father was one of his most avid readers—I had dreamed of getting to know him better. Like Clarke's, my own career has been driven by the imagination and wonder that comes from looking up at the stars, and I have also wanted to communicate that wonder. Quite surreptitiously, I had the chance to do that when he and I were asked to appear together on a PBS program the week Comet Shoemaker-Levy 9 collided with Jupiter. As we saw in chapter 19, that was a week to remember: The Shoemakers and I were in Washington; and the comsat system Clarke proposed connected him from his home in Sri Lanka to viewers across the United States. As we talked, people with small telescopes were looking at huge black clouds on Jupiter that might have been taken from one of his movies—but this time it was real. This time it was science and science fiction rolled into one.

Based on that first meeting, we began a sort of e-mail friendship, trading occasional notes and ideas. Our e-mail notes are still brief and concise, but I treasure each one of them. As I sit at my home in Arizona, looking out toward my observatory, Clarke sits in his wheelchair in Sri Lanka, writing about the future and how he hopes we'll get to it. It's a relationship Dad would have been proud of.

What is this future that Clarke so optimistically portrays? For Dad, that future was revealed through books like *The Exploration of Space*, which Clarke published in 1951 in the United Kingdom and in 1952 in the United States.[3] For me, though, it was clearly the movie *2001: A Space Odyssey*, which Stanley Kubrick directed and for which Clarke wrote the novel and screenplay. The film became extremely important in defining the philosophy of the United States as a spacefaring nation, and to me as an astronomy student having trouble passing college math. "David," mother said when I came home for Christmas vacation at the end of 1968, "you've got to go see *2001*. Its subject is nothing less than humanity's origins, and its future."

LIFE AND TIMES OF A VISIONARY

2001 is part of the cosmic vision of Arthur C. Clarke, arguably the greatest science fiction writer since Jules Verne. (Clarke disagrees with that assessment, insisting that "H. G. Wells is greater than us both!"[4]) Born in Somerset, England, in 1917, Clarke discovered the world of science fiction when he was eleven. As a teenager he built a transmitter that used a beam of light to send speech across short distances. During World War II Clarke was a radar officer in the Royal Air Force. He helped run its Ground Control Approach system, invented by American scientist Luis Alvarez, and spent some time at a radio school not far from the ancient druid monument Stonehenge, a place that thousands of years ago might have seen use as an observatory.

At the close of the war, the German scientists who had built and launched their V2 rockets scattered to the Soviet Union or the United States. On the gypsum sands of the New Mexico desert, they worked with American scientists to rebuild and launch these rockets. The British knew about the power of rockets in war as the V2s had hit London; at the White Sands Proving Grounds, scientists were learning about the possibilities of rockets in peace. Rocket power, as seen by the German V2s, was becoming a major new force. People like Pluto discoverer Clyde Tombaugh (see chapters 1 and 14), who at that time developed the optical tracking system for White Sands's launches, talked with Clarke about these possibilities. "Clyde helped smuggle this Brit spy," Clarke recalls that visit, "into White Sands!"[5] Decades later Tombaugh told me of those times when everyone was looking toward the stars. Back then, the promise of space was a tremendous thing.

Clarke saw the promise of space travel, but he also understood that governments would be loath to embrace it unless a practical use could be found. In October 1945, *Wireless World* published his letter proposing geosynchronous satellites. At that altitude, a satellite's orbital speed would match the Earth's rotation; the satellite would appear to hang over a single spot. A series of satellites in "Clarke orbits" could be used for communications. Clarke's idea was ignored at first, as the United States and the Soviet Union struggled for superiority in the postwar years. Then, on October 4, 1957, the Soviet Union launched *Sputnik*, the world's first artificial satellite, into orbit around the Earth. American thinking changed overnight.

The United States was catapulted into the space race, science education improved by leaps, and Clarke's idea for geosynchronous satellites was taken seriously. "Communications and astronautics were inextricably entangled in my mind," Clarke explains, "with results that now seem inevitable."[6] Every time you look at a weather satellite picture or use satellite television, you are making use of a system of satellites in orbit 22,300 miles above the Earth.

"If I had not proposed the idea of geostationary relays," Clarke added with unusual modesty, "half a dozen other people would have quickly done so. I suspect that my disclosure may have advanced the cause of space communications by approximately fifteen minutes."[7] Hardly! Clarke, as a science fiction writer, could see this potential and suggested it first. More important, he wouldn't let the idea drop. In 1947 he wrote *Prelude to Space*, in which he pushed his idea for communications satellites further.[8] It was a novel set thirty years in the future, he envisioned a world where these satellites played an important role. It turned out to be prophetic indeed. Clarke has taken the idea a big leap forward in his recent novel *3001: The Final Odyssey*, the fourth in his *Odyssey* series. Where now many satellites serve us in Clarke orbits, a thousand years in our future Clarke sees most of humanity living in a giant geosynchronous ring, hanging 22,300 miles above the Earth and linked with three gigantic towers.

AN OPTIMISTIC FUTURE

Through a lifetime Arthur C. Clarke's work has always pointed to an optimistic future for humanity, a world in which we have found creative solutions to our problems and in which anything is possible. In fact, Clarke has a law, which he first stated in his 1962 *Profiles of the Future*, voicing that optimism clearly: The only way to find the limits of the possible is by going beyond them to the impossible.[9]

Stretching the possible was certainly in Clarke's mind when he proposed a network of communication satellites almost two decades before the first one was launched. "I have reason to believe," Clarke wrote me, "that it had some influence on the men who turned this dream into reality. In the 22 years between the writing of *Prelude* and an actual landing on the Moon, our world changed almost beyond recognition. Back in 1947, I did not believe a lunar landing would be achieved even by that distant date. I would never have dared to imagine that by 1977, a dozen men would have walked on the Moon, and 27 would have orbited it."

Like many of us brought up under the wonder and hope of the space program, Clarke is still saddened by the decision, made long ago, to abandon the Moon as a goal for exploration. My father loved reading Clarke's solidly written stories about our future in space. I grew up on those books, and on *2001*. I remember the fervent discussion the movie generated about our future in space.

Where is that future now? "Still less could I have imagined," Clarke concluded, "that the first nation to reach the Moon would so swiftly abandon it again." It is only in the last few years that our nation has begun to focus its space program again. With an international station growing in space, and several probes headed toward other worlds, we are once again at a crossroads. The opportunity exists for some visionary to come and point the way—a task that Clarke has done so well for more than fifty years.

LOOKING INTO SPACE
AND PREDICTING THE FUTURE

The title of this segment seems to hark back to astrology, the practice of predicting our future from the stars. In Clarke's case, the predictions are based not on tables and tradition but on the enthusiasm of a brilliant man who looks thoughtfully toward the stars. Those predictions reached most people not through his books but through the film *2001* which appeared at the end of 1968, a year that brings a shudder to those of us who remember that defining moment in American history. The Communist Tet offensive in Vietnam was followed by the assassinations of Martin Luther King and Robert Kennedy, and the violence surrounding the Democratic National Convention. By any measure, 1968 was horrible. But Christmas eve that year, *2001* was attracting viewers with Clarke's message of optimism for humanity's future. "I believe that if you are an optimist," Clarke e-mailed me, "you have a chance of creating a self-fulfilling prophecy."[10] Just as Clarke tried to do exactly that with *2001*, three real-life astronauts circled the Moon aboard *Apollo 8* for the first time, sending greetings that same Christmas eve "to all of you, all of you on the good Earth."

2001: A Space Odyssey begins at the very dawn of civilization, as two prehuman tribes struggle for scarce water and food in Africa. Their lives are profoundly changed with the sudden appearance of a big stone slab some fifteen feet high that looked like a model of the United Nations building, only totally black, right in front of their rocky home. Acting as a sort of teacher, the monolith inspired one of the tribes to develop tools. In one of the movie's most memorable scenes, a protohuman, euphoric over his discovery of using a bone as a tool, throws it into the air. As it hurtles into the sky, we suddenly find its motion is slowed. Now twirling in slow motion through the air, the bone suddenly transforms into a futuristic space station gracefully turning in space. A second and larger monolith, it turns out, has been dug up on the Moon, and has sent a signal to Jupiter. Now it is December 2001, and astronauts heading to Jupiter aboard a ship called *Discovery* begin to learn about a cosmos where intelligence abounds, and where somehow, one of these intelligences comes back to help us along.

2001 offered a vision of where the world would be at the beginning of the new millennium. In the movie, people talked to one another over videophones, as a graceful, twirling space station orbited the Earth. Inside, passengers could dine at a Howard Johnson's or fly a Pan Am shuttle back to Earth. Though Howard Johnson's and Pan Am are not what they once were, the concept still stands. There was at least one permanent base on the Moon, at the crater Clavius close to the Moon's south pole. Clavius, perhaps not incidentally, is just north of another crater that straddles the Moon'south pole. Now named Shoemaker, that crater marks the spot where *Lunar Prospector* carried Gene Shoemaker's ashes to the Moon.

It seemed in the movie that by the year 2001, travel at least to orbit around the Earth should be as common as flying across the Atlantic. There would be at least one computer, named Hal, with artificial intelligence. Hal not only processed information, made calculations, and talked, but also *thought*. Hal not only ran the Jupiter ship but engaged its crew in chess, and—when confronted with contradictory instructions from its programmers—tried to take over the ship and change the destiny of its mission. Computers may not have advanced quite as far as that, not yet, but who would have imagined that by 2001, computers would be almost as common as telephones?

The real 2001 *seems* different. After twelve Americans walked on the Moon and explored that world, the U.S. government made a decision to abandon our nearest neighbor in space. "NASA had serious plans to land on Mars," Clarke recalled that time, "by the 1980s!"[11]

The real 2001 may *seem* different, but the variations between what Kubrick and Clarke foresaw and what we have now are almost entirely in the details. In fact, a videophone like the one he depicted in the film was actually used during that Shoemaker-Levy-Clarke PBS television show in 1994. We do not have a base on the Moon, but a space station is rapidly coming together high above the Earth. It's not a hotel yet, but the potential is there! "Virtually everything shown in the movie," Clarke wrote, "can be achieved during the next few decades. The time is right for the DC3 of space"[12]—just as that old aircraft was the workhorse of aviation two generations ago, our grandchildren might honeymoon or fly into space routinely aboard a space plane. Clarke won't apologize for his futuristic picture of our world three decades ago. In essence, he had it right.

WHERE ARE WE HEADING?

Thankfully, we still have the benefit of Arthur C. Clarke's insight. At age eighty-four, Clarke still lives in Sri Lanka, a place he moved to during the 1950s for its excellent diving, and because he had suffered through "too many English winters." He contemplates the future from a large, book-lined room in a place called Cinnamon Gardens; "a spice odyssey," he teases; "One of my windows looks out on my extensive garden; the other on the blank wall of a ladies college!"[13]

3001: The Final Odyssey is the last of Clarke's four *Odyssey* books. In it, Clarke boldly stretches his vision not thirty years ahead, but a thousand. If humanity proceeds in its present direction, and manages to survive, where will we be? I was both amazed and frightened by his answer. Much of what we do now for our work, mental exercise, or recreation, involves the telephone, cell phones, computer, and television. In Clarke's Earth of 3001, all these devices will have merged into a single unit called a Braincap. All humans will wear one, carefully adjusted to interact with the mind of its host. They would not be installed surgically. "I hope," says Clarke, that "physical contact on a bald skull will be sufficient, via nanoelectrodes."[14] Braincaps will allow us to communicate, participate in interactive movies, do business with one another—just about everything you can imagine, without using interfaces like a keyboard or telephone.

What a way to live! I asked Clarke if such technology could be used by the government or industry to control our minds. The technology itself, he replied, is neutral. "Of course they can!" he quipped; how we use this power is up to us. Another idea: now that civilization is truly global in 3001, the whole planet follows a universal time—whatever time it is naturally at the Greenwich Observatory in England is the time everywhere. The "confusing patch of time zones had been swept away," wrote Clarke, "in the advent of global communications."[15] That might work fine in theory, but how would New Yorkers like watching the Sun set at midnight after a fine summer day, or rise at noon in winter? We have no idea where we will be, or how we will live, a thousand years from now. Humanity—if we exist at all then—can go in a thousand different directions. But it is part of the fun and wonder of science and technology to imagine how we will live that far ahead.

As I finish reading the latest e-mail missive from Clarke, daylight has all but faded from the sky. Far around the world, his large garden must now be shrouded in predawn darkness, just as my desert yard will soon be dark. I shut down the computer and head outside to my observatory, pull open its sliding roof, remove the covers from the telescope, and point it skyward. I prepare to examine a few stars, some galaxies, and maybe a comet or two. But before I look through the scope, a big, brilliant moving light grabs my attention. It's the ISS—the International Space Station—which for the first time, at the dawn of 2001, is large enough to rival Jupiter in brightness. For me, that distant space station, made real by *2001: A Space Odyssey*, is the inspiration of Arthur C. Clarke who, in this real way, has brought space to Earth and made it a part of our lives and our dreams.

EPILOGUE

I stood and stared; the sky was lit,
The sky was stars all over it,
I stood, I knew not why,
Without a wish, without a will,
I stood upon that silent hill
And stared into the sky until
My eyes were blind with stars and still
I stared into the sky.
                    —Ralph Hodgson, 1913[1]

W hen Wendee and I began this book, there was no lack of suggestions
as to which discoverers should be included. We had no intention of
making this a complete guide to astronomical discovery, which would be impossible, since really anyone who has looked up at the sky, it could be said, has discovered something. We wanted to present a cross-section of people who,
throughout astronomical history, have lived with the hopes for and the consequences of their discoveries.

One consequence most people do not know of is "what will you do next?"
Clyde Tombaugh, for example, discovered Pluto when he was twenty-four years old.
At the time, his youth did not serve him well. Some of the other astronomers grew
jealous of Tombaugh's fame, a feeling that persisted long after he left Lowell Observatory in 1945, even though, as we have seen in chapter 14, Tombaugh went on to
make other important discoveries and complete some fascinating work. In the public
eye, he was known for Pluto and nothing else. "A professional famous person" was
what one planetarium director called him. Tombaugh was swept along on this tidal

wave of public opinion, which crashed ashore at the end of his life during the controversy over whether his planet should be demoted to the status of an asteroid.

The truth about Tombaugh is that he was a brilliant man who would probably have been a highly regarded scientist even if Pluto had never entered his life. His studies of Mars, his many discoveries of clusters, a dwarf nova, and a supercluster of galaxies were significant contributions to humanity's understanding of the cosmos.

Gene Shoemaker's lifetime of accomplishment also came with a price. His research was so well known that, as his fame grew, he, incredibly, found it almost impossible to obtain funding for his search for asteroids and comets at Palomar. In fact, other than a small account from the U.S. Geological Survey that covered transportation and housing during the observing periods, there was no special funding at all for the project until we discovered Comet Shoemaker-Levy 9. As a result of that find, NASA agreed to pay for film expenses only for an additional year. An additional frustration was that other searchers were being funded for doing essentially the same work. Although the Shoemakers did receive funding for the first few years of their fieldwork in Australia, for the last few years even that was cut off, and their fieldwork continued out of the couple's own pockets. As an employee of the U.S. government, Shoemaker was also prohibited from accepting fees for any of his public lectures.

Despite these setbacks, neither Tombaugh nor the Shoemakers were particularly unhappy about these problems. They merely accepted their situation as part of the price that discovery demands. Like Tennyson's Ulysses, they understood that while earthly matters of funding, politics, friendships, and other aspects of their lives come and go, the night sky is always there, waiting.

As this book goes to press, Carolyn Shoemaker continues her search for comets with the two authors of this book. In our "Shoemaker-Levy Double Cometograph" program, we are using an eight-inch and a twelve-inch Schmidt camera to take picture after picture, over and over again, in a survey for new comets. Although we have a lot of experience, we're still fooled by strategically placed defects in the film emulsion, or by ghost images of bright stars that masquerade as comets. But the search continues every bit as enthusiastically as it did when I first started in 1965, when Carolyn started in 1980, and when Wendee began in 1995. What keeps the search going, more than anything else, is that feeling of anticipation that comes each evening when the Sun goes down in a sky so clear that it beckons us to search, one more time, through its hidden corners for a new world:

> for my purpose holds
> To sail beyond the sunset, and the baths
> Of all the western stars, until I die . . .
> To strive, to seek, to find, and not to yield.

NOTES

CHAPTER 1: 1842: TO SAIL BEYOND THE SUNSET

1. Alfred, Lord Tennyson, "Ulysses," in *Victorian Poetry and Poetics*, ed. Walter Houghton and G. Robert Stange (Boston: Houghton Mifflin, 1968), p. 31.

2. John Milton, *Areopatigica*, in *John Milton: Paradise Lost and Selected Poetry and Prose*, edtied by Northrop Frye (New York: Holt, Rinehart and Winston, 1951, 1967), p. 487. See also Stillman Drake, *Essays on Galileo and the History and Philosophy of Science* (Toronto: University of Toronto Press, 1999), pp. 244–45.

3. Drake, *Essays on Galileo*, p. 288.

4. John Milton, *Paradise Lost I*, ll. 286–91; in Frye, *John Milton*, p. 13.

5. John Milton, *Paradise Lost V*, ll. 257–66; in Frye, *John Milton*, p. 115.

6. James Reston, *Galileo: A Life* (New York: HarperCollins, 1994), p. 279.

7. Arthur C. Clarke, *The Exploration of Space* (New York: Harper and Brothers, 1951), p. 191.

8. Clyde Tombaugh, interview with David Levy, 1985. Also see David H. Levy, *Clyde Tombaugh: Discoverer of Planet Pluto* (Tucson: University of Arizona Press, 1991), p. 61.

9. Clyde Tombaugh list provided by Patricia E. Tombaugh and used with her kind permission. I have tried to reproduced the list exactly as Tombaugh wrote it.

CHAPTER 2: 1965: PASSPORT FOR DISCOVERY

1. Leslie C. Peltier, *Starlight Nights: The Adventures of a Stargazer* (1965; rpt. Cambridge, Mass.: Sky Publishing Corporation, 1999), p. 231.

2. CN-3 Record Book, 1965.

3. Brian Marsden, "The Comet Pair 1988e and 1988g," unpublished paper.

4. Donald K. Yeomans, *Comets: A Chronological History of Observation, Science, Myth, and Folklore* (New York: John Wiley and Sons, 1991), p. 410.

CHAPTER 3: 1572: TYCHO BRAHE

1. Sarah Williams, "The Old Astronomer," in *Twilight Hours: A Legacy of Verse* (London: Strahan and Co., 1869), pp. 68–71.

2. Henry More, *Philosophicall Poems* (Cambridge: Printed by Roger Daniel, Printer to the University, 1647), p. 167.

3 Joseph Ashbrook, "Tycho Brahe's Nose," *Sky & Telescope* (June 1965): 353–54. See also Joseph Ashbrook, *The Astronomical Scrapbook* (Cambridge, Mass.: Sky Publishing Corporation and Cambridge, England: Cambridge University Press, 1984), pp. 3–5.

4. *Sky & Telescope* (October 1944): 12. The position is updated to 1950 coordinates.

5. See Owen Gingerich, "Tycho Brahe and the Great Comet of 1577," *Sky & Telescope* (December 1977): 452–58.

6. William Shakespeare, *King Henry IV, Part Two*, II, iv, 253–57.

7. Owen Gingerich, "Great Conjunctions, Tycho, and Shakespeare," *Sky & Telescope* (May 1981): 394–95 expands on this interesting relation of the sky and literature.

8. Ibid., p. 395.

9. A description of Tycho's observatory, especially Stjerneborg, appears in Erik Simonsen, "A Visit to Tycho Brahe's Observatory," *Sky & Telescope* (February 1974): 86–88.

10. *Oxford English Dictionary* suggests that the first such use was in Edward Blount's 1600 translation of a work by Conestaggio about the end of the House of Portugall. See G. Conestaggio, *The historie of the uniting of the kingdom of Portugall to the crown of Castill,* trans. E. Blount (London: A. Hatfield, 1600).

11. Fred Hoyle, *Astronomy* (New York: Doubleday, 1962), p. 106.

12. More, *Philosophicall Poems*, pp. 389–90.

13. Arthur Berry, *A Short History of Astronomy* (1898; New York: Dover, 1961), p. 140.

14. C. M. Huffer, "The Astronomy of Tycho Brahe," *Sky & Telescope* (January 1947): 11.

15. J. L. E. Dreyer, *Tycho Brahe: A Picture of Scientific Life and Work in the Sixteenth Century* (Edinburgh: A. and C. Black, 1890), p. 309.

16. Ashbrook, "Tycho Brahe's Nose," p. 354.

17. William Shakespeare, *Julius Caesar,* I, ii, 140–41.

CHAPTER 4: 1610: GALILEO AND THE INTERPRETATION OF DISCOVERY

1. Stillman Drake, trans. *Dialogue Concerning the Two Chief World Systems* (Berkeley: University of California Press, 1953, 1967). Einstein Foreward translated by Sonja Bargmann.

2. Stillman Drake, *Essays on Galileo and the History and Philosophy of Science,* Vol. 2 (Toronto: University of Toronto Press, 1999), p. 127.

3. Stillman Drake, *Essays on Galileo,* Vol. 1, 18.

4. James Reston Jr., *Galileo: A Life* (New York: HarperCollins, 1994), p. 52.

5. Drake, *Essays on Galileo,* Vol. 1, pp. 7–8.

6. Dava Sobel, *Galileo's Daughter* (New York, Penguin, 2000), pp. 4–5.

7. Arthur Berry, *A Short History of Astronomy* (1898; New York: Dover, 1961), pp. 149–50.

8. Then official Hague records were first generally published in 1831 by G. Moll, *Journal of the Royal Institution,* 324. See Henry C. King, *The History of the Telescope* (1955; New York: Dover, 1979), p. 31.

9. Galileo, *Sidereus Nuncius (The Starry Messanger)* in *Discoveries and Opinions of Galileo,* ed. and trans. Stillman Drake (New York: Doubleday Anchor, 1957), pp. 51–53.

10. Joseph Ashbrook, *The Astronomical Scrapbook* (Cambridge, Mass.: Sky Publishing Corporation and Cambridge, England: Cambridge University Press, 1984), pp. 198–99.

CHAPTER 5: 1632: GALILEO AND THE CONSEQUENCE OF DISCOVERY

1. Arthur Berry, *A Short History of Astronomy from Earliest Times through the Nineteenth Century* (1898; New York: Dover, 1961), p. 170.

2. Giorgio de Santillana, *The Crime of Galileo* (Chicago: University of Chicago Press, 1955, rpt. 1976), p. 310.

3. Stillman Drake and C. D. O'Malley, trans. *The Controversy on the Comets of 1618* (Philadelphia: University of Pennsylvania Press, 1960), p. 7.

4. Ibid., p. 19. The thought from Horace is from *Carmina* I, i, 36.

5. Ibid., pp. xxi–xxv.

6. Galileo, "The Assayer," 1623; Drake and O'Malley, *Controversy,* p. 164.

7. de Santillana, *The Crime of Galileo,* pp. 187–88.

8. Ibid., p. 157.

9. Ibid., p. 191.

10. Galileo, *Dialogue Concerning the Two Chief World Systems,* trans. Stillman Drake (Berkeley: University of California Press, 1953, 1967), p. 276.

11. Stillman Drake, "Reexamining Galileo's Dialogue," in *Essays on Galileo and the History and Philosophy of Science,* Vol. 2 (Toronto: University of Toronto Press, 1999), pp. 54–55.

12. Owen Gingerich, "How Galileo Changed the Rules of Science," *Sky & Telescope* (March 1993): 34.

13. James Reston, *Galileo: A Life* (New York: HarperCollins, 1994), p. 237.

14. de Santillana, *The Crime of Galileo*, p. 207.

15. Reston, *Galileo: A Life*, p. 243.

16. Dava Sobel, *Galileo's Daughter* (New York: Penguin, 2000), p. 243.

17. Stillman Drake, "On the Conflicting Documents of Galileo's Trial," in *Essays on Galileo*, p. 151.

18. de Santillana, *The Crime of Galileo*, p. 298.

19. Sobel, *Galileo's Daughter*, p. 279.

20. Reston, *Galileo: A Life*, p. 278.

21. de Santillana, *The Crime of Galileo*, p. 329.

22. Gingerich, "How Galileo Changed the Rules," p. 36.

CHAPTER 6: 1656: CHRISTIAAN HUYGENS

1. John Donne, "The First Anniversary," ll. 205–18, in *The Poems of John Donne,* ed. Sir Robert Grierson (London: Oxford University Press, 1933), vol. 1, pp. 237–38.

2. Carl Sagan, *Cosmos* (New York: Random House, 1980), p. 143.

3. Louis Bell, *The Telescope* (1922; New York: Dover, 1981), pp. 16–17.

4. William Sheehan, *Worlds in the Sky: Planetary Discovery from Earliest Times through Voyager and Magellan* (Tucson: University of Arizona Press, 1992), p. 132.

5. Ibid., p. 133.

6. Henry C. King, *The History of the Telescope* (1955; New York: Dover, 1979), p. 54.

7. Sheehan, *Worlds in the Sky*, p. 123.

CHAPTER 7: 1682: EDMOND HALLEY

1. Leslie C. Peltier, *Starlight Nights: The Adventures of a Stargazer* (1965; Cambridge, Mass.: Sky Publishing Corporation, 1999), pp. 16–17.

2. Donald K. Yeomans, *Comets* (New York: John Wiley & Sons, 1991), p. 115.

3. Ibid., pp. 112–13.

4. Ibid., p. 113.

5. See Dava Sobel, *Longitude: The True Story of a Lone Genius Who Solved the Greatest Scientific Problem of His Time* (New York: Penguin, 1995).

6. Yeomans, *Comets*, p. 122.

7. Brian G. Marsden to David Levy, October 1, 1992.

8. Edmond Halley, *Astronomical Tables, 1749, 1752*; see Yeomans, *Comets*, p. 122.

9. Yeomans, *Comets*, pp. 122–23.

10. Nicolas-Louis de Lacaille, *Mémoires de mathématique et de physique, tirés des registres de l'Académie Royale des Sciences, de l'année 1759*; see also Yeomans, *Comets*, p. 138.

CHAPTER 8: 1760: CHARLES MESSIER

1. Steven James O'Meara, *Deep Sky Companions: The Messier Objects* (Cambridge, Mass.: Sky Publishing Corporation and Cambridge, England: Cambridge University Press, 1998), p. ix.

2. Kenneth Glyn Jones, *Messier's Nebulae and Star Clusters*, 2d ed. (Cambridge: Cambridge University Press, 1991), p. 344.

3. Ibid., p. 347.

4. Charles Messier, *Connaissance des Temps*, 1809; see Jones, *Messier's Nebulae*, p. 351.

5. Jones, *Messier's Nebulae*, p. 353.

6. Peltier, *Starlight Nights*, p. 228.

7. Yeomans, *Comets*, pp. 150–51.

8. Ibid., p. 151.

9. Sobel, *Longitude*, p. 54.

10. Although this story appears in several sources, a particularly good one is Jones, *Messier's Nebulae*, p. 365. Some of the details of Messier's relationship with de Saron come from this source as well.

11. Ibid., pp. 364–65.

12. Ibid., pp. 367.

13. David H. Levy, "Charles Messier and His Catalogue," in O'Meara, *Deep Sky Companions*, p. 5.

CHAPTER 9: 1781: CAROLINE AND WILLIAM HERSCHEL

1. W. Herschel announced his discovery at the end of March 1781 to the Bath Literary and Philosophical Society. This "Account of a Comet" appears in *The Scientific Papers of Sir William Herschel*, ed. J. L. E. Dreyer (London: The Royal Society and the Royal Astronomical Society, 1912), 1: 30–38.

2. C. A. Lubbock, *The Herschel Chronicle* (Cambridge: Cambridge University Press, 1933), p. 15.

3. Ibid., p. 60.

4. Ibid., p. 66.

5. W. G. Hoyt, *Planets X and Pluto* (Tucson: University of Arizona Press, 1980), p. 12.

6. Charles Messier to William Herschel, in Lubbock, *The Herschel Chronicle*, p. 86.

7. Lubbock, *The Herschel Chronicle*.

8. Henry C. King, *The History of the Telescope* (1955; New York: Dover, 1979), p. 124.

9. W. Herschel to C. Herschel, July 3, 1782. See also "America's Last King and His Observatory," in Joseph Ashbrook, *The Astronomical Scrapbook* (Cambridge, Mass.: Sky Publishing Corporation, 1984), p. 17.

10. Ashbrook, "William Herschel and the Sun," *Astronomical Scrapbook*, pp. 338–39.

11. Ibid., p. 338.

12. Lubbock, *The Herschel Chronicle*, p. 246.

13. Gary W. Kronk, *Comets* (Hillside, N.J.: Enslow, 1984), p. 264.

14. N. A. Mackenzie, "He Broke Through the Barriers of the Skies," *Sky & Telescope* 8, no. 5 (1949): 119. The title is a translation of the words on Herschel's tombstone, *Coelorum perrupit claustra*.

15. King, *The History of the Telescope*, p. 133.

16. Ibid., p. 128.

17. Ibid., p. 142.

18. Hoyt, *Planets X and Pluto*, p. 15.

19. Ashbrook, "John Herschel's Expedition to South Africa," in *Astronomical Scrapbook*, pp. 38–39.

CHAPTER 10: 1846: ADAMS, LEVERRIER, AND THE SCANDAL OVER NEPTUNE

1. W. M. Smart, "John Couch Adams and the Discovery of Neptune," *Occasional Notes of the Royal Astronomical Society* 2, no. 11 (1947): 47. See also W. G. Hoyt, *Planets X and Pluto* (Tucson: University of Arizona Press, 1980), p. 40.

2. Alfred, Lord Tennyson, "In Memoriam" (XXI, lines 17–20), in *Victorian Poetry and Poetics*, ed. Walter E. Houghton and G. Robert Stange (Boston: Houghton Mifflin Co., 1968), p. 51.

3. John Couch Adams, "An Explanation of the observed Irregularities in the Motion of Uranus, on the hypothesis of disturbances caused by a more distant planet; with a determination of the mass, orbit and position of the disturbing body," *Appendix to the Nautical Almanac for the Year 1851* (London: 1846), p. 3. See also Hoyt, *Planets X and Pluto*, p. 40.

4. George Airy, "Account of some circumstances historically connected with the discovery of the Planet exterior to Uranus," *Monthly Notices of the Royal Astronomical Society* 7 (1846): 123. See also Hoyt, *Planets X and Pluto*, p. 42.

5. Hoyt, *Planets X and Pluto*.

6. Airy, "Account of Some Circumstances."

7. Joseph Ashbrook, "The Airy Regime at Greenwich," in *The Astronomical Scrapbook* (Cambridge, Mass.: Sky Publishing Corporation, 1984), p. 44.

8. H. H. Turner, obituary of J. G. Galle, *Monthly Notices of the Royal Astronomical Society* 71 (1911): 278. See also Hoyt, *Planets X and Pluto*, p. 52.

9. M. Grosser, *The Discovery of Neptune* (New York: Dover, 1979), p. 117.

10. Ibid., p. 119.

11. Ibid., p. 128.

12. Airy, "Account of Some Circumstances," p. 143. See also Hoyt, *Planets X and Pluto*, p. 54, and Grosser, *The Discovery of Neptune*, pp. 129, 131.

13. Adams, "Explanation," p. 5. See also Hoyt, *Planets X and Pluto*, p. 57.

CHAPTER 11: 1912: HENRIETTA LEAVITT, HARLOW SHAPLEY, AND OUR PLACE IN THE MILKY WAY GALAXY

1. Letter to E. C. Pickering, March 1, 1905, in Bessie Zaban Jones and Lyle Gifford Boyd, *The Harvard College Observatory: The First Four Directorships, 1839–1919.* (Cambridge: The Belknap Press of Harvard University Press, 1971), p. 367.

2. Ibid., p. 400.

3. Ibid.

4. David H. Levy, *Observing Variable Stars: A Guide for the Beginner* (Cambridge: Cambridge University Press, 1989.)

5. Katherine Haramundanis, *Cecilia Payne-Gaposchkin: An Autobiography and Other Recollections* (Cambridge: Cambridge University Press, 1984), p. 207.

6. E. C. Pickering to H. N. Russell, November 7, 1917, in David H. DeVorkin, *Henry Norris Russell: Dean of American Astronomers* (Princeton, N.J.: Princeton University Press, 2000), p. 149.

7. Jones and Boyd, *Harvard College Observatory*, p. 367.

8. Henrietta Swan Leavitt, "1777 Variables in the Magellanic Clouds," *Annals of the Astronomical Observatory of Harvard College* 60, part IV (1908): 107.

9. "Periods of 25 Variable Stars in the Small Magellanic Cloud," *Harvard Circular* 173 (March 3, 1912). See Harlow Shapley, ed., *Source Book in Astronomy: 1900–1950* (Cambridge, Mass.: Harvard University Press, 1960), p. 188.

10. Leavitt, "1777 Variables," pp. 186–87.

11. Haramundanis, *Cecilia Payne-Gaposchkin*, p. 146.

12. Ibid.

13. Mildred Shapley Matthews, interview with David H. Levy, December 12, 1990.

14. Harlow Shapley, *Through Rugged Ways to the Stars* (New York: Charles Scribner's Sons, 1969), pp. 52–53.

15. Haramundanis, *Cecilia Payne-Gaposchkin*, p. 153.

CHAPTER 12: 1919: EDDINGTON, EINSTEIN, MERCURY, AND AN ECLIPSE

1. Abraham Pais, *Subtle Is the Lord: The Science and the Life of Albert Einstein* (Oxford: Clarendon Press, Oxford University Press, 1982), p. 253.

2. Albert Einstein, "Does the Inertia of a Body Depend upon Its Energy Content?" *Annalen der Physik* 18 (1905): 639–41.

3. Pais, *Subtle Is the Lord*, p. 253.

4. A. Vibert Douglas, *The Life of Arthur Stanley Eddington* (London: Thomas Nelson and Sons, 1956), p. 2.

5. Arthur Stanley Eddington, *Notebook*, see Douglas, *Life*, p. 13.

6. Douglas, *Life*, p, 39.

7. Sir Arthur Eddington, *Space, Time, and Gravitation: An Outline of the General Relativity Theory* (Cambridge: Cambridge University Press, 1920), p. 113.

8. Eddington, *Notebook*, see Douglas, *Life*, p. 40.

9. Eddington, *Space, Time, and Gravitation*, pp. 114–15.

10. Ibid.

11. Douglas, *Life*, p. 40.

12. Eddington, *Space, Time, and Gravitation*, p. 117.

13. Douglas, *Life*, p. 41.

14. Ibid., pp. 43–44.

15. Ibid., p. 60.

16. Ibid., p. 163.

CHAPTER 13: 1924: EDWIN HUBBLE

1. Gale E. Christianson, *Edwin Hubble, Mariner of the Nebulae* (Chicago: University of Chicago Press, 1995), p. 158.

2. Ibid., pp. 46–47.

3. Ibid., p. 72.

4. David H. Levy, *The Man Who Sold the Milky Way: A Biography of Bart Bok* (Tucson: University of Arizona Press, 1993), p. 44.

5. Christianson, *Edwin Hubble*, p. 90.

6. Ibid., p. 158.

7. Harlow Shapley, *Galaxies* (Cambridge, Mass.: Harvard University Press, 1943).

8. Bart J. Bok, interview with David H. Levy, July 31, 1982.

9. Bart J. Bok, "The Apparent Clustering of External Galaxies," *Harvard Bulletin* No. 895 (1934): 1–8.

10. Gary W. Kronk, *Comets* (Hillsdie, N.J.: Enslow, 1984), pp. 130–31.

11. Christianson, *Edwin Hubble*, p. 357.

12. Ibid., p. 358.

CHAPTER 14: 1930: CLYDE TOMBAUGH

1. William Shakespeare, *Twelfth Night*, II, v, 131–34.

2. Albert G. Ingalls, "The Heavens Declare the Glory of God," *Scientific American* (November 1925).

3. Clyde Tombaugh, interview with David H. Levy, 1985.

4. V. M. Slipher to Clyde W. Tombaugh, November 30, 1928. Reprinted by permission.

5. Slipher to Tombaugh, December 21, 1928.

6. Slipher to Tombaugh, January 2, 1929.

7. William Graves Hoyt, *Planets X and Pluto* (Tucson: University of Arizona Press, 1980), p. 133.

8. Ibid., p. 164.

9. Clyde Tombaugh, lecture, July 1980, Astronomical Society of the Pacific.

10. Clyde Tombaugh, interview with David H. Levy, November 8, 1985.

11. Roger Lowell Putnam to Clyde Tombaugh, March 26, 1930. Reprinted by permission.

12. Chancellor of the University of Kansas to Clyde Tombaugh, March 21, 1930. Reprinted by permission.

13. Adella Chritton Tombaugh to Clyde Tombaugh, March 16, 1930. Reprinted by permission.

14. Muron Tombaugh to Clyde Tombaugh, March 16, 1930. Reprinted by permission.

15. Clyde Tombaugh, "Planet Discoverer Tells How He Did It," *Daily Science News Bulletin*, Science Service, March 16, 1930.

CHAPTER 15: 1948: BART BOK

1. David H. Levy, *The Man Who Sold the Milky Way: A Biography of Bart Bok* (Tucson: University of Arizona Press, 1993), p. 1.

2. Bart Bok, interview with David H. Levy, 1983.

3. Ibid.

4. M. Shapley Matthews, interview with David H. Levy, December 12, 1990.

5. Bessie Zaban Jones and Lyle Gifford Boyd, *The Harvard College Observatory: The First Four Directorships, 1839–1919* (Cambridge, Mass.: The Belknap Press of Harvard University Press, 1971), p. 191.

6. Ibid., p. 192.

7. David H. Levy, *Observing Variable Stars: A Guide for the Beginner* (Cambridge: Cambridge University Press, 1989), pp. 170–71.

8. Bart J. Bok, *Study of the Eta Carinae Region* (Cambridge, Mass.: Harvard College Observatory Reprint, No. 77, 1932).

9. Bart J Bok, *The Distribution of the Stars in Space* (Chicago: University of Chicago Press, 1937).

10. Bart J. Bok and Edith F. Reilly, "Small Dark Nebulae," *Astrophysical Journal* 105 (1947): 255.

11. Levy, *The Man Who Sold the Milky Way*, p. 64.

12. Walter Baade to Bart Bok, March 25, 1947.

13. Bok to Baade, April 7, 1947.

14. Fred Hoyle, *The Black Cloud* (New York: Signet, 1959), p. 18.

15. Ibid., p. 108.

16. Levy, *The Man Who Sold the Milky Way*, p. 67.

17. Jeff Hester to David H. Levy, November 1995.

18. E. J. Maggio to David H. Levy, 1983. See also Levy, *The Man Who Sold the Milky Way*, p. 186.

19. Bart J. Bok, "The Promise of the Space Telescope," *Congressional Record*, vol. 125, pt. 11 (1979): 13504–506.

CHAPTER 16: 1951: THE MILKY WAY IS A SPIRAL GALAXY

1. Carl Sagan, *Cosmos* (New York: Random House, 1980), p. 261.
2. Grote Reber, "Cosmic Static," *Astrophysical Journal* 91 (1940): 621–24.
3. Walter Baade to Bart Bok, February 8, 1949.
4. William Morgan, "The Spiral Structure of our Galaxy," in *The New Astronomy* (New York: Simon and Schuster, 1954), p. 95.
5. O. Struve, "Galactic Exploration by Radio," *Sky & Telescope* 11, no. 9 (1952): 215.
6. B. J. Bok, "Radio Studies of Interstellar Hydrogen," *Sky & Telescope* 13, no 12 (1954): 408.

CHAPTER 17: 1963: SEYFERT GALAXIES AND QUASARS

1. David H. Levy, *The Man Who Sold the Milky Way: A Biography of Bart Bok* (Tucson: University of Arizona Press, 1993), pp. 31–32.
2. Frederick Golden, *Quasars, Pulsars, and Black Holes* (New York: Scribner's, 1976, Pocket Books, 1977), p. 104.
3. Gerald Cecil, Jonathan Bland-Hawthorn, and Sylvain Veilleux, "The Lives of Quasars," in *The Scientific American Book of the Cosmos*, ed. David H. Levy (New York: St. Martin's Press, 2000), pp. 49–56.
4. *Sky & Telescope* 20, no. 2 (August 1960).
5. M. A. Stull, "Two Puzzling Objects: OJ 287 and BL Lacertae," *Sky & Telescope* 45 (1973): 224–26.

CHAPTER 18: 1979: USING A GALAXY AS A TELESCOPE

1. See Leif J. Robinson, "Giant Galactic Arcs," *Sky & Telescope* 73 (April 1987): 379.
2. "The Case of the Double Quasar," *Sky & Telescope* 58 (November 1979): 427.
3. Robinson, "Giant Galactic Arcs," p. 379.
4. Ibid.
5. "Giant Arcs: Light Echoes or Lensed Galaxies?" *Sky & Telescope* 75 (January 1988): 7.
6. "Arcs Galore," *Sky & Telescope* 76 (October 1988): 358.
7. "New Arcs and Rings," *Sky & Telescope* 86 (July 1993): 15.
8. "Merging Galaxies and Bright Arcs," *Sky & Telescope* 87 (January 1994): 11.

CHAPTER 19: 1993:
COMET SHOEMAKER-LEVY 9

1. Dylan Thomas, "Do Not Go Gentle into That Good Night," in *Miscellany One: Poems, Stories, Broadcasts* (London: J. M. Dent and Sons Ltd., 1963), p. 31.
2. One of the best technical reviews of the origin of life is J. N. Marcus and M. A. Olsen's paper "Biological Implications of Organic Compounds in Comets," in R. L. Newburn et al., *Comets in the Post-Halley Era* (Dordrecht: Kluwer Academic Publishers, 1991), pp. 439–62.
3. J. R. Hind, *The Comet of 1556; Being popular replies to every-day questions, referring to its anticipated reappearance, with some observations on the apprehension of danger from comets* (London: John W. Parker and Son, 1857), p. 45.
4. See Tom Gehrels, ed. *Hazards Due to Comets and Asteroids* (Tucson: University of Arizona Press, 1994).
5. David H. Levy, *The Man Who Sold the Milky Way: A Biography of Bart Bok* (Tucson: University of Arizona Press, 1993), cover, facing p. 109.
6. David H. Levy, *The Quest for Comets: An Explosive Trail of Beauty and Danger* (New York: Plenum, 1994).
7. H. Weaver to David H. Levy, 23 February 1995.

CHAPTER 20: 1999: FINDING PLANETS
AROUND OTHER STARS

1. Geoffrey Marcy to David Levy, January 13, 2001.
2. Brad Smith to David Levy, November 21, 1988. See also Clyde Tombaugh et al., *The Search for Small, Natural Earth Satellites: Final Technical Report* (Las Cruces: New Mexico State University Physical Science Laboratory, 1959), p. 66.
3. Paul Butler to David Levy, January 12, 2001.
4. Ibid.
5. Ken Croswell, *Planet Quest: The Epic Discovery of Alien Solar Systems* (New York: The Free Press, 1997), pp. 140–45.
6. Dale Frail to David Levy, July 21–22, 2001.
7. Geoffrey Marcy to David Levy, January 13, 2001.
8. Ibid.
9. Ibid.
10. American Astronomical Society meeting, January 2001, San Diego.
11. T. Brown et al., *Astrophysical Journal Letters* 23 (November 1999).
12. A list of the nearest stars is found in Rajiv Gupta, *Observer's Handbook* (Toronto: Royal Astronomical Society of Canada, 2001), pp. 234–35.
13. Geoffrey Marcy and R. Paul Butler, "The Diversity of Planetary Systems," *Sky & Telescope* 95, no. 3 (March 1998): 36–37.
14. Clyde Tombaugh, conversation with David H. Levy, summer 1996.

15. See Peter D. Ward and Donald Brownlee, *Rare Earth: Why Complex Life Is Uncommon in the Universe* (New York: Springer-Verlag, 1999).

16. Philip Morrison, ed., *The Search for Extraterrestrial Intelligence* (NASA SP-419; New York: Dover, 1979).

17. The SETI@home Sky Survey, see http://www.setiathome.ssl.berkeley.edu/sciencepaper.html. See also Sullivan et al., "A New Major SETI Project Based on Project SERENDIP Data and 100,000 Personal Computers," in *Astronomical and Biochemical Origins and the Search for Life in the Universe, Proceedings of the Fifth International Conference on Bioastronomy, IAU Colloquium no. 161,* ed. Batalli Cosmovici, Stuart Bowyer, and Dan Werthimer (Bologna, Italy: Editrice Compositori, 1997).

CHAPTER 21: 1971: ACQUAINTED WITH THE NIGHT

1. Robert Frost, "Acquainted with the Night," in *Robert Frost: Collected Poems, Prose, and Plays,* ed. Richard Poirier and Mark Richardson (New York: Library of America, 1995), p. 234.

2. Thomas Hardy, *Far from the Madding Crowd* (New York: Harper and Brothers, 1912), p. 9.

3. Brad Smith, *Voyager* Imaging Team Leader, presentation at Flandrau Planetarium, 1979.

4. Stephen James O'Meara to David Levy, January 8, 2001.

5. *International Astronomical Union Circular* 3192, February 7, 1984.

CHAPTER 22: 1988: JEAN MUELLER

1. Michael A. Seeds, *Horizons* (Belmont, Calif.: Wadsworth Publishing, 1981), p. 158.

2. Jean Meuller to David H. Levy, February 1992.

3. Alfred Noyes, *The Torchbearers: Watchers of the Sky* (New York: Frederick A. Stokes Co., 1922), p. 2.

4. Jean Mueller to David Levy, February 1992.

CHAPTER 23: 2001: TO DISCOVER AN IDEA

1. Jules Bergman, ABC News, summer 1962.

2. Stuart J. Weidenschilling et al., "Photometric Geodesy of Main-Belt Asteroids, I: Lightcurves of 26 Large, Rapid Rotators," *Icarus* 70 (1987): 191–245. See also *Icarus* 76 (1988): 19–77 and *Icarus* 86 (1990): 402–47.

3. Arthur C. Clarke, *The Exploration of Space* (New York: Harper and Row, 1952).

4. Arthur C. Clarke to David Levy, 14 January, 2001.

5. Ibid.

6. Clarke to Levy, November 2000.

7. Ibid.

8. Arthur C. Clarke, *Prelude to Space* (London: Sidwick and Jackson, 1953).

9. Arthur C. Clarke, *Profiles of the Future* (New York: Harper and Row, 1962).

10. Clarke to Levy, December 15, 2000.

11. Clarke to Levy, November 2000.

12. Clarke to Levy, January 3, 2001.

13. Clarke to Levy, September 29, 2000.

14. Clarke to Levy, January 14, 2001.

15. Arthur C. Clarke, *3001: The Final Odyssey* (New York: Ballantine Books, 1997), p. 25.

EPILOGUE

1. Ralph Hodgson, *Collected Poems* (London: Macmillan, 1961).

INDEX